THE
PRACTICAL
GEOLOGIST

THE PRACTICAL GEOLOGIST

Dougal Dixon

Raymond L. Bernor, Editor

A FIRESIDE BOOK
Published by Simon & Schuster Inc.
New York London Toronto Sydney Tokyo Singapore

A QUARTO BOOK

Simon and Schuster/Fireside
Simon and Schuster Building
Rockefeller Center
1230 Avenue of the Americas
New York, N.Y. 10020

Copyright © 1992 by Quarto Inc

Designed and produced by Quarto
Publishing plc,
The Old Brewery, 6 Blundell Street,
London N7 9BH

Senior Editor Caroline Beattie
Editor Richard Jones
Designer Nick Clark

Picture Research Manager Sarah
Risley
Illustrators Dave Kemp, Janoś
Marffy, Rob Shone
Art Director Moira Clinch
Publishing Director Janet Slingsby
Typeset by Bookworm Typesetting
Manufactured by J Film Process
Singapore Pte Ltd
Printed by Leefung-Asco Printers Ltd China

4 5 6 7 8 9 10 Pbk

Dixon, Dougal.
    The practical geologist/Dougal
Dixon.
      p.   cm.
    Includes index.
    ISBN 0–671–74698–7 (cloth). —
ISBN 0–671–74697–9 (Fireside pbk)
    1. Geology–Guide-books.   I. Title.
QE46.D69 1992
550 — dc20

∪–671–74698–7
0–671–74697–9 Pbk   91–23914
                                              CIP

CONTENTS

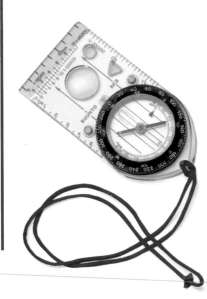

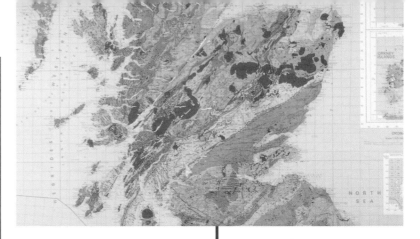

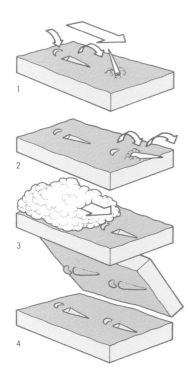

ROCK COLLECTION

NAME: PLAGIOCLASE FELDSPAR

WHAT IS A PRACTICAL GEOLOGIST?

Geology – the science of the Earth. It is a study that incorporates all the other sciences and binds them together in one all-embracing subject.

Literature gives us some guides. In *Saint Ronan's Well* by Sir Walter Scott, written in 1823, Meg Dodds, the prickly landlady of Cleikum Inn, refers to those of her guests who:

> *"rin uphill and down dale, knapping the chucky stanes to pieces wi' hammers, like sae mony road-makers run daft, to see how the world was made."*

In *A Study in Scarlet* (1887), Sir Arthur Conan Doyle, speaking through Dr. Watson, lists Sherlock Holmes' accomplishments and his limitations. In the list we read:

> *"Knowledge of Geology – Practical, but limited. Tells at a glance different soils from each other.*

Below A landscape immediately presents the practical geologist with many questions. Some of these are obvious, and can be answered immediately. Others require some background knowledge and will stimulate field investigations.

Why is there farmland here, and not here?

Why is this ridge so different from the rest of the landscape?

Are these houses built of local rocks? If so, what are they?

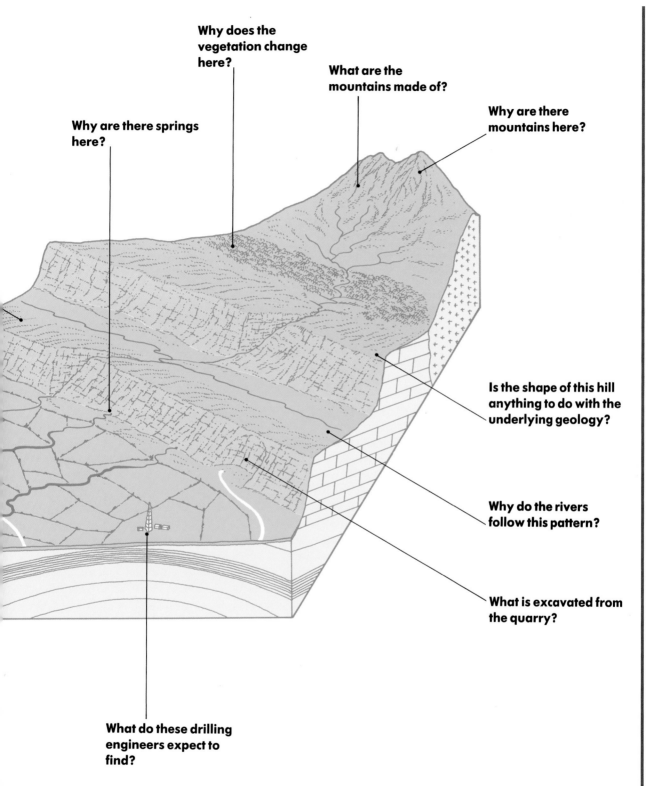

Why does the vegetation change here?

What are the mountains made of?

Why are there mountains here?

Why are there springs here?

Is the shape of this hill anything to do with the underlying geology?

Why do the rivers follow this pattern?

What is excavated from the quarry?

What do these drilling engineers expect to find?

After walks has shown me splashes upon his trousers, and told me by their colour and consistence in what part of London he had received them."

Here we have two aspects of the science of geology – the academic and the utilitarian. The former represents knowledge for its own sake, while the latter knowledge is turned to some creative purpose. Both aspects involve practical work – going out to find the knowledge in the first place.

Observational geology such as this has a long history. Greek scholars such as Pythagoras (*c* 580–500 BC) and Herodotus (*c* 485–425 BC) both noted the presence of fossil seashells high up in mountains and drew the conclusion that geographies were very different in times past. This early surge of interest in geology vanished during the Dark Ages and did not surface again in the West until the Renaissance in the 15th century. At this time technology and the arts began to blossom, and the necessity of supplying the raw materials for these new activities led to an interest in the formation of minerals. In 1556 the German mineralogist Georgius Agricola published *De Re Metallica* in which he describes the formation of metal ores in veins in a manner that was well ahead of his time.

The observations made over the next century or two led to some erroneous theories. The presence of crystals in some igneous rocks suggested to Abraham Gottlob Werner (1749–1817), Professor of Mining and Mineralogy at Freiberg, that all rocks had been deposited from solution as a vast primordial ocean had evaporated. This view – the Neptunian view – became geological orthodoxy.

The value of fieldwork
Throughout history the observations of the Earth's structure and composition have been misinterpreted, and it is only by diligent field work and increasingly precise experimentation that the more realistic theories have been developed.

One of the pioneers of utilitarian geology was English canal engineer William Smith (1769–1839). During the course of his work he realized that the different layers of rocks through which he excavated his ditches and tunnels could be identified by the kinds of fossils that they contained. Using this information, he was able to construct the first geological map and, with his book *Strata Identified by Organized Fossils*, initiated the science of stratigraphy.

Nowadays the classic study of geology has combined with such related subjects as meteorology, oceanography, astronomy, geophysics, and geochemistry, to become the all-embracing discipline of Earth Science. For the professionals, the practical work is done with highly sophisticated equipment. Drills penetrate the surface of the Earth and bore rock samples from deep within the crust. The structure of the underground rocks can be studied by setting off explosions and recording the patterns of reflected shock waves, analyzing

James Hutton

In the late 1700s Scotland became the home of practical geology. When we look at a geological map of the world it is not difficult to see why. Scotland, within its 49 000 square miles (80 000 square km), possesses practically every geological structure and age of rock possible. The person regarded as the founder of modern geology is James Hutton (1726–97), of Edinburgh. By studying the rocks where they outcropped (in the field, as geologists say), he formulated theories about the past conditions that formed them. He visualized an abiding Earth on which forces of rock formation the results by computer. Infra-red photography from aircraft and satellites can show chemical differences in vegetation that reveal the nature of the underlying rocks. Sonar waves bounced off the ocean floor can give resolute pictures of the landscape. Sensitive gauges can measure the electrical properties of soil and rocks, and determine if an exploitable water supply lies beneath. Instruments can detect tiny variations in the Earth's gravitational field which can suggest the presence of workable metal deposits. It is all a long way from hammers and mud stains.

With the great surge of 20th-century knowledge and the wealth of books on Earth Science – at both the academic and the popular level – it has become possible to learn all that is known about the nature and the workings of the Earth without moving from one's armchair. Yet practical geology for the amateur is far from dead. Studying the rocks as they outcrop and collecting new specimens will always add to the mounting knowledge of the Earth. And nothing can quite compare with the exhilaration of tramping up a deserted mountainside to observe the folds and faults in the exposed rock, or with the joy of splitting open a boulder to find a crystal of garnet or a trilobite fossil that has never before been exposed to the daylight.

Geology in Scotland

were balanced by forces of rock destruction. It was he who put the Neptunian theories to rest.

Another approach to practical geology is that of the experimental scientist. The pioneer in this field was Sir James Hall (1761–1832) – not the great 19th-century American geologist of the same name but another

Scot from Edinburgh. Hall melted rock specimens in the furnace of an iron foundry and observed what was produced as they cooled. He also conducted experiments to simulate the formation of various rock structures by compressing layers of clay. Most of these experiments were conducted after Hutton's

death, because he was dismissive of this experimental approach, scorning those who tried to "judge the great operations of the mineral kingdom by kindling a fire and looking into the bottom of a crucible." Since smelting furnaces are probably beyond the reach of most readers, somewhat simpler

experiments will be considered in this volume.

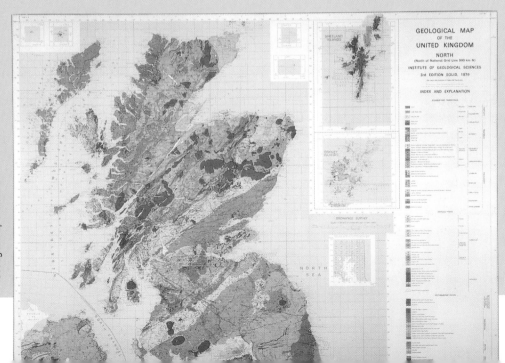

Geological map of Scotland

THE HISTORY OF THE EARTH

The Earth is made of stardust, as is all it contains, including ourselves. We cannot actually trace the Earth's history, but the following is agreed to be a rough approximation.

THE DEVELOPMENT OF THE SOLAR SYSTEM

Once upon a time, about 15 000 million years ago, there was a vast mass of dust and gas in one of the spiral arms of our galaxy, and this mass began to contract. The spaces between the stars are not empty; they contain gases and tiny dust particles in varying concentrations. The arms of the galaxy are not static, but spin and pulsate, with waves passing through them, although very slowly by our standards. A random concentration of gas and dust, caught up by one of these waves, would have been compressed to such an extent that the miniscule gravitational forces between the particles would come into operation, and the particles would have begun to move together. The original movement of the cloud in its orbit around the galaxy would have caused the contracting mass to begin to spin.

Two main forces then came into play. The first was the gravitational force between the particles, and this caused all the matter to compress toward the center where the greatest mass was accumulating. The second was a centrifugal force that spun the matter outward again along a plane perpendicular to the axis of rotation. As a result of both these forces the matter began to form in a broad spinning disk which was our Solar System in embryo.

The greatest mass of material gathered at the center, and the energy released by the collision of particles and the compression of the material caused the mass to heat up. This action would have taken only a few million years – a very short time in geological terms – and the Sun was "switched on."

The disk of matter was not stable. Eddies appeared across its face, disrupting the local speed of the spin. As a result, the material on the inner arms of the eddies was orbiting at a smaller angular velocity and tended to fall toward the "protosun" while that on the outer arms was orbiting faster and was thrown outward. The mass of the disk was thus separated into a number of discrete stable rings around the protosun, similar to the rings around some of our planets. Only a few hundred thousand years would have been needed to create this ring system – a system that would eventually become the nine planets of the Solar System.

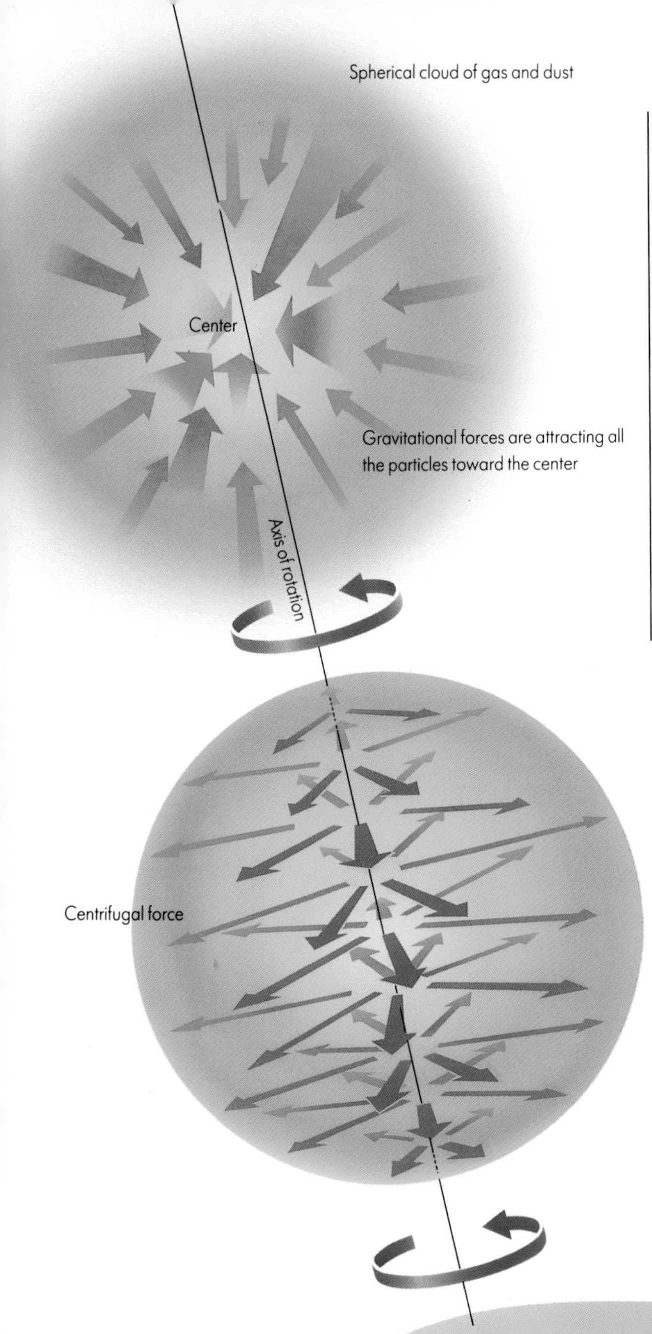

Spherical cloud of gas and dust

Center

Gravitational forces are attracting all
the particles toward the center

Axis of rotation

Something to do

Blow up a bicycle tire. As you build up the pressure, feel how the barrel of the pump becomes hot. This is because the energy that is given out by the air as it compresses must show itself in some way – in this case in the form of heat.

The formation of the protosun involved the generation of heat in this way. Normally the heat would radiate away into space and be lost. However, in the evolving Sun there was a build-up of gravitational force as well as compressional force, and much of the heat generated remained within the Sun and began the nuclear reactions that are continuing to this day.

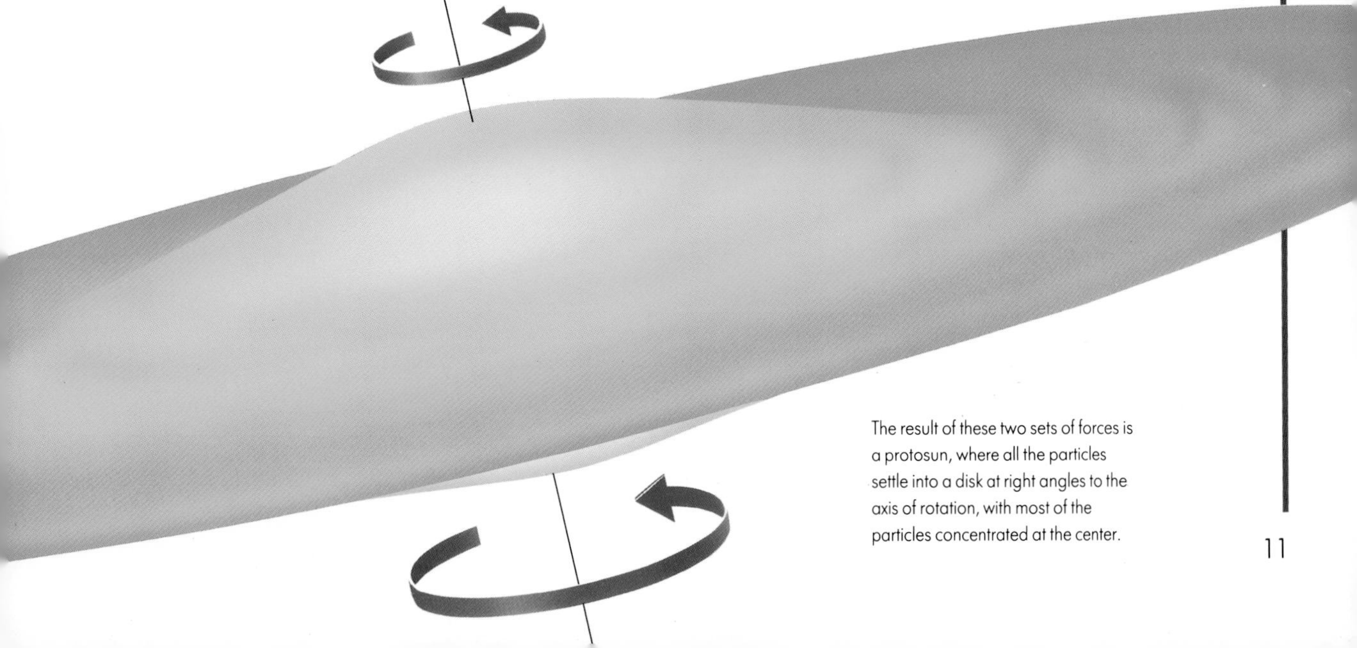

Centrifugal force

Since the cloud is spinning, centrifugal forces are throwing all the particles away from the axis of rotation

The result of these two sets of forces is a protosun, where all the particles settle into a disk at right angles to the axis of rotation, with most of the particles concentrated at the center.

11

THE EARTH SOLIDIFIES

The rings of gas and dust particles around the protosun were subjected to the same random wave effects as the galaxy arms, and in any area where the concentration became particularly great the matter began to gravitate together to form lumps, probably around 100 yards or meters in diameter. Eventually these lumps began to collide and stick together. The larger accumulations scooped up the smaller pieces as the rings revolved, and in each ring all the ring matter began to accumulate into a single large mass – a "planetesimal."

Let us now concentrate on the planetesimal that eventually became the Earth. There are two main theories about the process of accretion. According to the first – the "homogeneous accretion model" – all the particles accumulated in a random mass, with every component spread throughout the planetesimal without a pattern of any sort. The same action that generated the heat of the protosun then generated heat in the embryonic Earth. The heat melted the iron and nickel in the mass and the droplets of these, being heavy, sank toward the center. The stony material – the silicates – being lighter, would have remained on the outside.

The alternative theory – the "heterogeneous accretion theory" – suggests that the iron and nickel gathered together to form the first planetesimal while the silicate material was still drifting about as the remains of the ring. The silicate then settled on the outside.

Whatever the process, the result is that we now have an Earth that is divided into layers. There is an inner core of solid iron and nickel, an outer core of molten iron and nickel, a mantle of dense stony silicate material, and a crust of lighter silicate material.

We have a close neighbor in space, which provides more problems to scientists who try to decipher the early history of the Earth. That is the Moon. The Apollo astronauts of the late 1960s and early 1970s brought back rock samples that proved to be about four billion years old. That is about the age at which we think the Earth finally solidified, suggesting that the Earth and the Moon formed at the same time. But how? Again a number of theories have been proposed, none of which is absolutely accepted. The Moon seems to be nearly all mantle, with a considerably smaller iron-rich

Above From away out in space we can see little of the Earth but cloud and haze. It looks blue because of reflection from the atmosphere and the oceans. The shapes of the continents can be glimpsed through the cloud cover. Astronauts in a

low orbit can see the mountain ranges, the river courses, the island chains. The geologist on the ground can actually touch the Earth's substance, and make deductions about our planet's history.

core. It has been suggested that a lump of the Earth's mantle became wrenched away by some force before it solidified. Another theory suggests that the Moon was a planetesimal trapped in Earth's orbit while the whole Solar System was settling down. Or maybe it condensed from the cloud of mostly silicate particles that was slowly accreting around Earth's iron/nickel planetesimal (provided we accept the heterogeneous accretion theory).

Thereafter the Moon has had a considerably less exciting history than the Earth. During the first 600 million years or so it suffered an intense meteorite bombardment – presumably as the last of its component material was drawn to it. This produced the rugged crater-rich landscape that characterizes the light areas of the Moon's surface. Shortly after that it heated up, and molten material erupted and spilled across the surface, forming the dark patches called the maria (singular mare), formerly thought to be seas. All that was about three billion years ago. Nothing has happened to the Moon since, compared to Earth's checkered history.

Something to do

You can simulate the heterogeneous accretion theory by filling a transparent plastic cup with particles of different densities. Take a piece of Styrofoam, or expanded polystyrene, and tease it out into its individual bubbles. (The static electricity that then causes the bubbles to stick to your hands and tools can be thought of as analagous to the gravitational forces that caused the cloud particles to come together – but that is not the object of this exercise.) Put the results in the cup along with some small stones. Shake the entire concentration, and you will find that the heavy stones soon gather at the bottom.

The heterogeneous accretion theory can be demonstrated by putting the small stones in the bottom of the cup first, and then adding the Styrofoam.

INSIDE THE EARTH

The Earth is composed of layers with distinct physical and chemical properties. The chemical compositions are most easily compared to those of common minerals, which will be dealt with later in the book.

Continental crust

Oceanic crust

Upper mantle

Continental crust
- Outermost 16–56 miles (25–90 km) representing the continental masses.
- Composition extremely complex, but averaging that of granite. High in silica and aluminum – sometimes referred to as SIAL. More silica than oceanic crust.
- Density 195 lb/cubic ft (2500 kg/m^3), or about 2.5 times the density of water, but this depends greatly on the individual rocks.
- Temperature 1148–32°F (700–0°C).
- The whole crust represents only 0.7% of the Earth's mass.
- Much of this is speculation based on the results obtained by a number of different geophysical techniques that have been used to study the Earth's interior.

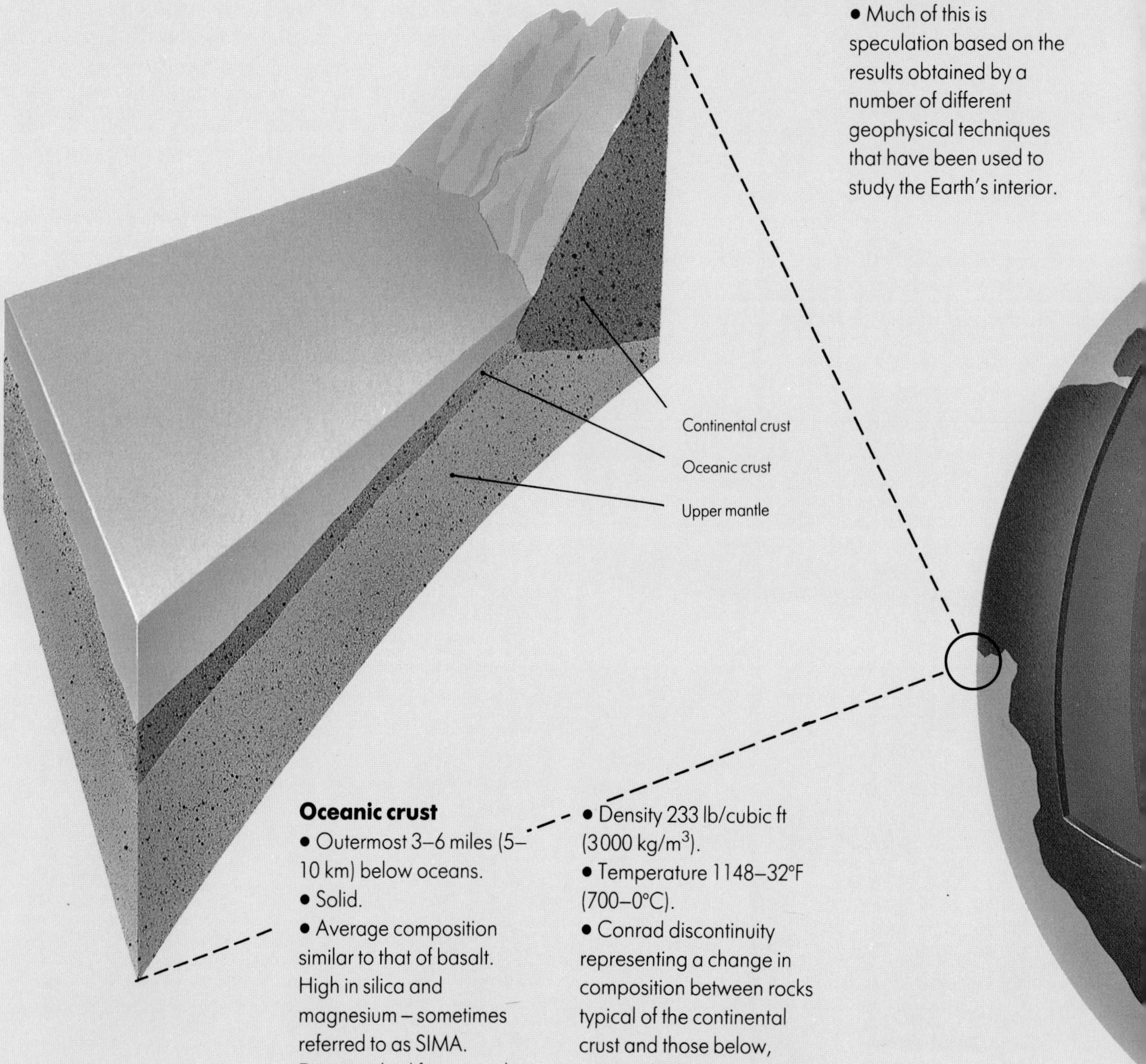

Oceanic crust
- Outermost 3–6 miles (5–10 km) below oceans.
- Solid.
- Average composition similar to that of basalt. High in silica and magnesium – sometimes referred to as SIMA. Distinguished from mantle composition by having more silica.
- Density 233 lb/cubic ft (3000 kg/m^3).
- Temperature 1148–32°F (700–0°C).
- Conrad discontinuity representing a change in composition between rocks typical of the continental crust and those below, intermediate in composition between ocean crust and upper continental crust.

Upper mantle

● 249 miles (400 km) to anything between 56 miles (90 km) beneath continents and 3 miles (5 km) beneath oceans.

● Solid except for a squashy layer near the outside, especially below the oceans, where some of the minerals are molten.

● Similar composition to rest of the mantle but richer in the mineral olivine.

● Recognizable minerals, such as spinel and garnet, sometimes extruded by volcanoes.

● Density 233 lb/cubic ft (3 000 kg/m^3).

● Temperature 2340–1148°F (1 300–700°C).

● Mohorovičić (Moho) discontinuity representing a change of composition.

Transition zone

● 652–249 miles (1 050–400 km).

● Solid.

● Similar composition to the rest of the mantle but showing changes between very compact mineral phases to looser, less dense phases.

● Density 358–268 lb/cubic ft (4 600–3 380 kg/m^3).

● Temperature 3 240–2 340°F (1 800–1 300°C).

Lower mantle

● 1793–652 miles (2 885–1 050 km).

● Solid.

● Composition similar to the minerals olivine (60%), pyroxene (30%), and feldspar (10%). Fairly even composition throughout.

● Density 420–358 lb/cubic ft (5 400–4 600 kg/m^3).

● Temperature 5 040–3 240°F (2 800–1 800°C).

● The whole mantle represents 68.3% of the Earth's mass.

Outer core

● 3762–1793 miles (5 155–2 885 km).

● Liquid. Moving about by convection and producing the Earth's magnetic field.

● Iron-sulfur mixture.

● Density 950 to 770 lb/cubic ft (12 200–9 900 kg/m^3).

● Temperature 5 760°F (3 200°C).

● 29.3% of the Earth's mass.

● Gutenberg discontinuity representing a great change in composition and also in density.

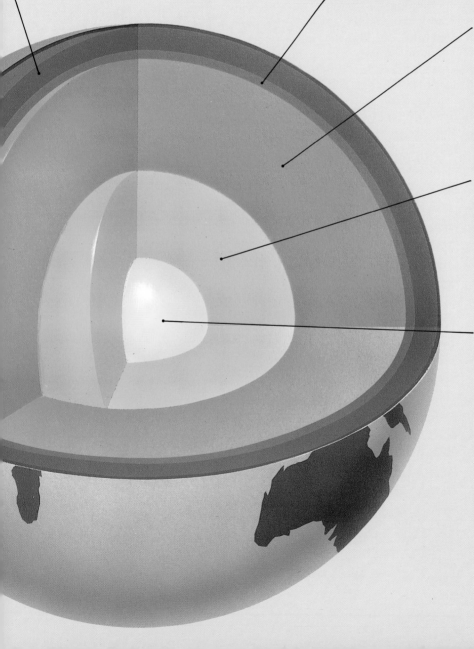

Inner core

Center of the Earth

● 3958–3762 miles (6370–5 155 km).

● Solid.

● Iron-nickel alloy.

● Density 1017 lb/cubic ft (13 000 kg/m^3), or about 13 times the density of water as we know it.

● Temperature 8 100°F (4 500°C).

● 1.7% of the Earth's mass.

● Discontinuity that shows a transition between a liquid and a solid phase. There is a sudden increase in density below it.

15

THE EVIDENCE FOR THE THEORIES

The interior of the Earth is as inaccessible to us as the interior of the Moon. Yet there are indirect ways of obtaining information about the subsurface features of our planet.

Meteoritic evidence

Particles of the nebula that spawned the Earth and the rest of the Solar System are still falling onto the surface of our planet. Occasionally these come in blocks large enough to survive the destructive heat of friction as they pass through the Earth's atmosphere, and so land as meteorites. Two types of meteorite are known: iron meteorites and stony meteorites. We can regard the iron meteorites as remains of the substances that formed the core of our planet, and the stony meteorites as representing the material of the mantle. The proportions of the two types are similar to the proportions of the core and mantle.

Eruptions of mantle material to the surface

This is a rare occurrence, but it gives us the opportunity of analyzing mantle material directly. Sometimes nodules of mantle appear in the basaltic lavas of oceanic volcanoes. The material for the basaltic lavas is extruded from the mantle but usually it changes completely through cooling and lowering of pressure before coming anywhere near the surface. The unaltered mantle nodules carried in it contain silica, but in a much smaller proportion than in crustal rocks. Other nodules, called peridotites, are found in a particular kind of ancient volcano, called a kimberlite pipe. These nodules are interesting from an economic viewpoint because they contain diamonds, formed at depths of 60 miles (150 km) or so.

Seismic surveys

The principle of studying the refraction of shock waves can be used on a local scale by generating artificial earthquakes. Controlled explosions are set off and the vibrations that are refracted through the various layers in the crust are re-

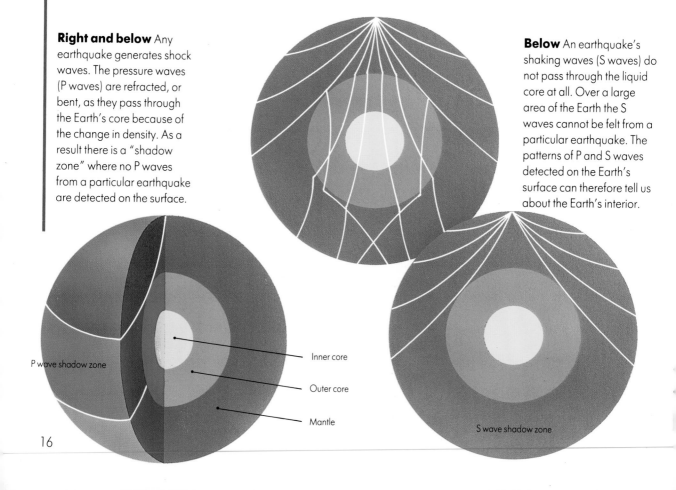

Right and below Any earthquake generates shock waves. The pressure waves (P waves) are refracted, or bent, as they pass through the Earth's core because of the change in density. As a result there is a "shadow zone" where no P waves from a particular earthquake are detected on the surface.

Below An earthquake's shaking waves (S waves) do not pass through the liquid core at all. Over a large area of the Earth the S waves cannot be felt from a particular earthquake. The patterns of P and S waves detected on the Earth's surface can therefore tell us about the Earth's interior.

P wave shadow zone

Inner core

Outer core

Mantle

S wave shadow zone

Left The Big Hole at Kimberley, South Africa, excavated in the diamond rush of the late 1800s, looks huge but is only 975 ft (300 m) deep – a pinprick in the Earth's crust. Direct study of the material of the planet's interior is impossible, but numerous physical and chemical techniques can be used in subsurface exploration.

corded by geophones and analyzed by computer. On a smaller scale the technique of seismic reflection (rather than seismic refraction) is used, and gives much finer results. In this the shock waves are bounced off the various layers within the crust. Nowadays the shock waves may be produced less violently but in a more controlled manner by vibrators mounted on road vehicles.

Drilling

The deepest hole drilled into the crust to date is in the Kola peninsula in the USSR. It has reached a depth of 7 miles (12 km) and the Soviet scientists plan to extend it to 9 miles (15 km). Even so, it is a mere pinprick on the Earth's surface. An international team, led by the United States, has been sailing the oceans since 1969 drilling into the much thinner ocean crust. No one has yet succeeded in reaching the Mohorovičić (Moho) discontinuity, the boundary between the crust and its underlying mantle.

Scientists can use the bore hole to discover several things. They can analyze the core samples that are brought up. Instruments can be lowered to test the electrical properties of the various layers penetrated. A sonic generator can be lowered into the hole to produce a sound source for acoustic logging – similar in principle to seismic refraction. Sensors can be lowered to record the differences in natural radioactivity between the layers. In practice all of these techniques are used in combination to produce as complete a picture as possible.

Gravity studies

The denser a material is, the greater will be its gravitational force. It takes a sensitive instrument to detect any difference in gravity between one place and another (a gravity anomaly), but it can be done. Variations in a satellite's orbit can be used to detect large-scale variations across the globe. Such large-scale variations are due to the fact that Earth is not a perfect sphere.

On the ground the anomalies can be measured using a plumb line. The plumb will be pulled minutely to one side in close proximity to a heavy mass. Sir George Everest found this in the mid-19th century when he was surveying India. On the northern plains of the Ganges and Indus his pendulum was pulled toward the high Himalayas to the north. Straightway he saw why, but on doing the measurements he found that the deflection was not as great as it should have been considering the mass of mountain there. Various interpretations were put on this, and even in the 1920s there were books that suggested that the particularly heavy soil of the Indo-Gangetic plain was pulling the pendulum back again, away from the mountains. The true interpretation, however, had been made in the 1850s by Sir George Airy, the then Astronomer Royal. He realized that the continental crust was thicker under mountains than elswhere, and penetrated more deeply into the mantle. Thus there was a great mass of light material to the north of Sir George Everest that was exerting a smaller pull than would be expected had the heavy mantle been close to the surface there – a negative gravity anomaly.

Such situations are routinely investigated using a device called a gravimeter. This consists of a weight suspended on a spring. A region of high gravity – a positive anomaly – will pull the weight and the spring farther downward than a negative anomaly, and the differences can be measured.

Accurately broad interpretations of the various geophysical results show that the crust beneath

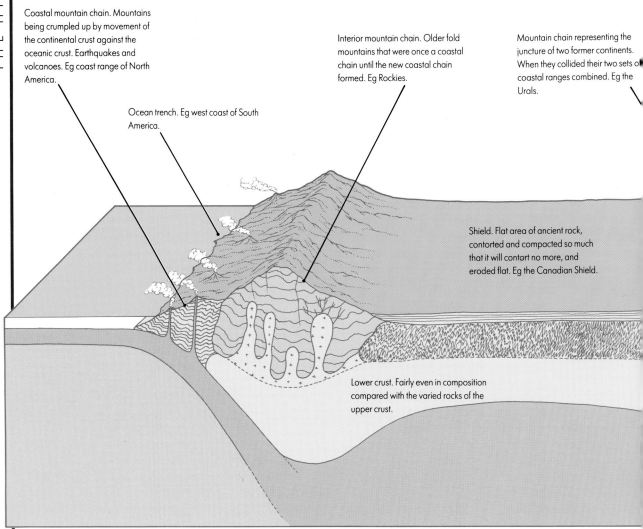

Coastal mountain chain. Mountains being crumpled up by movement of the continental crust against the oceanic crust. Earthquakes and volcanoes. Eg coast range of North America.

Interior mountain chain. Older fold mountains that were once a coastal chain until the new coastal chain formed. Eg Rockies.

Mountain chain representing the juncture of two former continents. When they collided their two sets of coastal ranges combined. Eg the Urals.

Ocean trench. Eg west coast of South America.

Shield. Flat area of ancient rock, contorted and compacted so much that it will contort no more, and eroded flat. Eg the Canadian Shield.

Lower crust. Fairly even in composition compared with the varied rocks of the upper crust.

the ocean is thin and dense. In the region of the continents, the crust is thick and light. A typical continent will consist of a solid contorted mass of extremely ancient rock at its heart – often called a shield – for example, the Canadian shield of North America. This will be surrounded by ranges of progressively younger mountains, such as the Rockies, and there may be very young mountains, such as the Coast Ranges, by the sea. Old mountains may lie across the continent, separating one shield area from another – for example, the Urals between Europe and Asia. In some areas

it may look as if the continent is cracking apart, as along the Great Rift Valley of East Africa. Some coasts, for example the Atlantic coast of South America, may give the appearance of having broken away from some other continental area. This is the broad picture of a continent – a structure built of rocks. And it is these rocks that will constitute the subject of the rest of this book.

Below A typical continent is made up of a number of parts – usually a core of ancient metamorphic rock surrounded by the remains of mountain ranges. Not all continents exhibit all the features shown here. The structure of each continent is a reflection of its history.

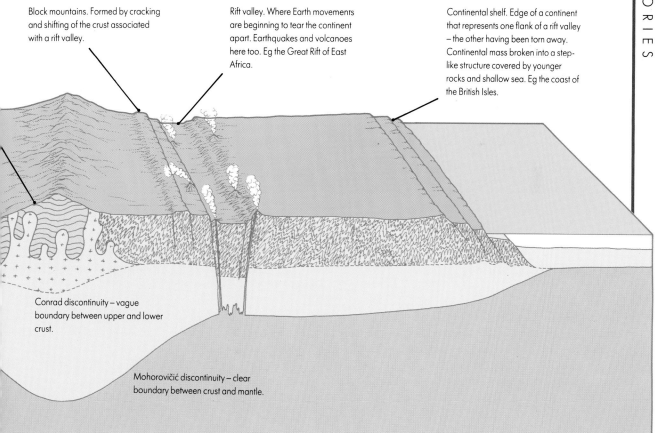

Block mountains. Formed by cracking and shifting of the crust associated with a rift valley.

Rift valley. Where Earth movements are beginning to tear the continent apart. Earthquakes and volcanoes here too. Eg the Great Rift of East Africa.

Continental shelf. Edge of a continent that represents one flank of a rift valley – the other having been torn away. Continental mass broken into a step-like structure covered by younger rocks and shallow sea. Eg the coast of the British Isles.

Conrad discontinuity – vague boundary between upper and lower crust.

Mohorovičić discontinuity – clear boundary between crust and mantle.

THE SUBSTANCE OF THE EARTH

Pick up a stone. In your hand you have a piece of the Earth's crust, made up of minerals produced by the chemical reactions that formed the planet.

MINERALS – ROCK COMPONENTS

If you take a rock, any rock, and look at it through a hand lens or a microscope you will see that it is made up of a mosaic of interlocking particles. Sometimes, in rocks such as granite, these are so big that you can see them with the naked eye. These particles are the minerals, naturally occurring homogeneous solids which have been inorganically formed. They have definite chemical compositions and definite atomic structures.

When a rock forms, the chemicals organize themselves into a number of different minerals. There are hundreds of different types, each with its own particular chemical composition, but some are more common than others. Every rock is made up of a mixture of different minerals – but usually no more than half a dozen or so.

For convenience we can divide the minerals into two broad classifications – the rock-forming minerals and the ore minerals. The latter are those that usually come to mind when the word mineral is mentioned, those that can be mined and processed for a product, but they are a very minor constituent of the Earth's crust.

As we have seen, silica (SiO_2) is the most common chemical component of the Earth, so the most common rock-forming minerals are silicates – minerals containing silica. Silica can take part in complex chemical reactions and so there are many different types of silicate mineral.

The simplest silicate mineral is quartz, which is pure silica. More commonly there are metallic elements combined with the silica. Magnesium forms a high proportion of the oceanic crust, and so the magnesium-iron silicate mineral called olivine ($(Mg,Fe)_2SiO_4$) is common here. Continental crust is rich in aluminum, and so continental rocks tend to be rich in the aluminum silicate minerals called the feldspars, such as orthoclase ($KAlSi_3O_8$) and albite ($NaAlSi_3O_8$).

Carbonates – compounds containing carbon – are also important rock-forming minerals. Perhaps the most important is calcite ($CaCO_3$). This tends to be unstable when exposed to the weather and so rocks containing large proportions of carbonate tend to be eroded more quickly than those containing the more robust silicate minerals.

The silicates contain metals. However, their chemical nature is such that the metals are almost impossible to remove. Olivine therefore cannot be regarded as a source of magnesium, any more than feldspar would be a useful storehouse of aluminum. Ore minerals must contain a metal that is easily extracted. Sometimes a mineral contains the metal and nothing else. Such "native ores" include gold nuggets.

Oxides – the metal combined with oxygen – are important ore minerals. Most of the iron ores are oxides, such as magnetite (Fe_3O_4) and hematite (Fe_2O_3).

A metal combined with sulfur forms a sulfide mineral, many of which are ore minerals. These include iron pyrites (FeS_2) and the lead mineral galena (PbS).

Left When we look closely at a rock we can see that it is made up of much smaller components. Sometimes they form good crystal shapes, and sometimes irregular chunks. These components are called the minerals.

Atoms and molecules

The figures in parentheses are the chemical formulas of the minerals. The smallest indivisible part of any chemical substance is the molecule, and a molecule is made up of atoms of various elements. In chemistry an element can be referred to by a symbol, usually the initial of its common name, such as S for sulfur, or, if there are more than one with the same initial, an abbreviation of its Latin name, such as Pb standing for *plumbum* – Latin for lead.

The chemical formula reflects the number of different atoms making up the molecule. Hence the chemical formula for quartz is SiO_2, which shows that each molecule consists of an atom of silica (Si) bonded to two atoms of oxygen (O). Other elements which are referred to in this book are iron (Fe), aluminum (Al), sodium (Na), potassium (K), calcium (Ca), copper (Cu), hydrogen (H), fluorine (F), berylium (Be), zircon (Z), magnesium (Mg), manganese (Mn), chromium (Cr), titanium (Ti).

Above Some minerals are economically important. These are called the ore minerals. Iron ore, for example, comes in several mineral forms, including the yellowish powdery limonite **top** and the unevenly shaped kidney ore **above**.

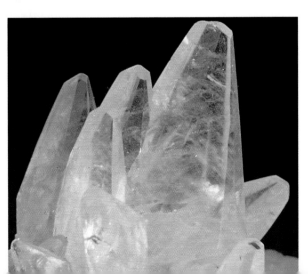

Left Most minerals, however, are referred to as the rock-forming minerals, since they make up the bulk of the rocks. Calcite is a common rock-forming mineral, but only rarely does it produce the well-shaped crystals seen here.

MINERAL SHAPES

If a mineral is allowed to grow unhindered it will develop a characteristic three-dimensional shape, a crystal.

The crystal form reflects the arrangement of atoms in the molecule. When quartz accumulates, the molecules of silica will be positioned upon one another only in a way that the shape of the molecule permits. Thus, many millions of molecules will build up in a regular pattern which will show itself in the shape of the final crystal.

The crystal shape is an important clue to the identification of a particular mineral, and geologists recognize six different crystal systems. Each is based on the number of axes of symmetry developed. An axis is an imaginary line running through the crystal around which it can be turned to produce the same appearance from more than one side.

The sides of a crystal are termed faces, and these meet one another in interfacial angles. This gives us a valuable rule in crystallography. A crystal hardly ever grows evenly. Sometimes one face grows faster than another and so the final crystal looks nothing like the theoretical type for that mineral. However, the law of constancy of interfacial angles states that angles between the faces are always the same no matter how distorted the crystal may be. The angles can be measured on

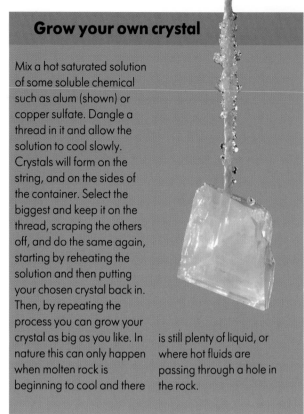

Grow your own crystal

Mix a hot saturated solution of some soluble chemical such as alum (shown) or copper sulfate. Dangle a thread in it and allow the solution to cool slowly. Crystals will form on the string, and on the sides of the container. Select the biggest and keep it on the thread, scraping the others off, and do the same again, starting by reheating the solution and then putting your chosen crystal back in. Then, by repeating the process you can grow your crystal as big as you like. In nature this can only happen when molten rock is beginning to cool and there is still plenty of liquid, or where hot fluids are passing through a hole in the rock.

Build your own goniometer

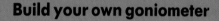

If you have a large crystal on which to practice you can make a simple contact goniometer from a semicircle of stiff cardboard. Mark out a gradation as shown on the diagram and pivot a pointer at the center. The angle between the faces can be read off directly.

More sophisticated goniometers mount the crystal on a turntable and reflect light off the faces, measuring the angle through which the crystal must be turned to bring a second reflection into the position of the first.

a device called a goniometer.

Related to the crystal structure is a property called the cleavage. Planes of weakness in the crystal lattice reveal themselves in the tendency for the crystal to split in a certain direction. A mineral like mica $(KAl_2(AlSi_3)O_{10}(OH,F)_2)$, in which the silicate molecules are arranged in flat sheets, can flake away like the leaves of a book. Others, such as calcite $(CaCO_3)$, have more than one cleavage plane, and upon shattering fragment into perfect mini-crystals.

To complicate matters, a crystal may grow in two different directions from one face. The result is called a twin. Twinned crystals can be recognized by the presence of re-entrant angles – something not found in single crystals.

Unfortunately for the geologist, minerals hardly ever form good crystals. When a rock forms, all the chemical components organize themselves into minerals, which grow crammed up against one another. Only if a mineral can develop in a fluid, uncluttered by other solid matter, does a good crystal form develop.

The mineral systems

The following systems are used to classify crystals.

Cubic system The crystal has three axes, all at right angles to one another, and all the same length. Iron pyrites (FeS_2) crystallizes in the cubic system.

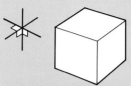

Tetragonal system The crystal has three axes, all at right angles to one another, two of which have the same length. Chalcopyrite – copper iron sulfide – $(CuFeS_2)$ has a tetragonal crystal.

Hexagonal system The crystal has four axes, three are of equal length, at 120° to one another and at right angles to the fourth. An example is the gemstone beryl – berylium silicate – $(Be_3Al_2Si_6O_{18})$ or $Be_3Al_2(SiO_3)6$.

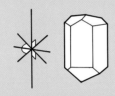

Orthorhombic system The crystal has three axes, all at right angles to one another, but of unequal lengths. Topaz – fluorine and aluminum silicate – $(Al_2F_2SiO_4)$ crystallizes in the orthorhombic system.

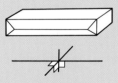

Monoclinic system The crystal has three axes, one not at right angles to the others, of unequal lengths. Augite – an iron and magnesium silicate – $(Ca,Mg,Fe,Al)_2(Al,Si)_2O_6)$ has a monoclinic crystal.

Triclinic system The crystal has three axes, none of which is at right angles to either of the others, and are all different lengths. Albite feldspar $(NaAlSi_3O_8)$ is an example.

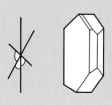

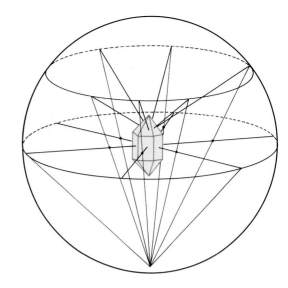

The faces of a crystal can be plotted on a plane. The mathematics are complex but the theory is simple. Imagine a crystal placed at the center of a sphere **above**. Project lines from the center at right angles to each face to meet the surface of the sphere. Connect each of these surface points to the south pole of the sphere. The pattern formed where these lines cut the equatorial plane of the sphere is the stereographic projection. It does not matter how distorted the crystal is, the resulting stereographic projection is the same, because of the constancy

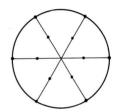

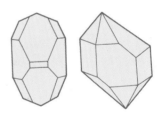

of interfacial angles.

In practice stereographic projections are plotted on a kind of circular graph called a Wulff net.

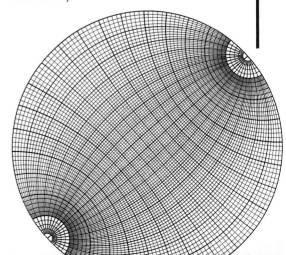

MINERAL IDENTIFICATION

A mineral sample that can be held in the hand and studied is called, naturally enough, a hand specimen. If a specimen does not have a distinctive crystal shape to reveal its identity, there are a number of other techniques that we can apply in order to identify it.

Hardness

Different minerals have different hardnesses, and these are measured on Mohs' scale. The scale runs:

1 Talc
2 Gypsum
3 Calcite
4 Fluorite
5 Apatite
6 Orthoclase
7 Quartz
8 Topaz
9 Corundum
10 Diamond

This is rather an arbitrary scale based on nothing more than the fact that a mineral at one point can scratch the surface of a mineral at a lower point, and will itself be scratched by a mineral at a higher point. It is not really practical for the amateur to possess a whole set of these minerals, but some everyday items of known hardness can be used instead:

2.5 Fingernail
4 "Copper" coin
5 Glass
5.5 Penknife blade
6.5 Steel file

Streak

Do not be misled by a mineral's color. Too often the color is determined by impurities, or by the freshness of the specimen. However, if the mineral is drawn over an unglazed white ceramic tile it will leave a streak of powder which is often a reliable guide to its composition. Find a piece of such a tile (the edge of a broken coffee cup is just

Left A mineral's fracture can be diagnostic. Conchoidal fracture, such as in this obsidian, produces patterns of concentric ridges.

Below Some minerals can be identified by their luster, such as the metallic sheen on these crystals of pyrites.

as good) and keep it with your mineralogy rig. Hematite, an iron ore mineral, will produce a cherry red streak, while pyrites will give a greenish or brownish-black streak.

Specific gravity

This is the weight of a mineral compared with the weight of an equal volume of water. The specimen is usually weighed in air. Then it is immersed in water and the water displaced is caught and weighed. Dividing the weight of the mineral by that of the water gives the value for the specific gravity. This is a somewhat cumbersome procedure, but you will, with practice, get a good idea of the specific gravity by hefting the specimen in your hand. Anything with a specific gravity of more than 3, such as fluorite, would be noticeable and something like galena, with a specific gravity of 7.6, would be very obvious.

Fracture

Minerals with distinct cleavage planes will split along them when broken. Others will break randomly, while some may give a distinctive type of break or fracture.

Conchoidal fracture is quite notable, producing concentric patterns of ridges reminiscent of some seashells. Quartz shows a conchoidal fracture.

Hackly fracture is a jagged surface as produced in some metals. Native metals usually have a hackly fracture.

Earthy fracture gives a powdery appearance. Meerschaum, a magnesium silicate, shows an earthy fracture.

Left Silky luster, such as in this specimen of gypsum, is typical of minerals that are formed of thick masses of very fine hair-like crystals.

Above Resinous luster, as seen in feldspar, can give the impression that the mineral is made of plastic. The way a mineral reflects light is important in identification.

Luster

The fresh surface of a mineral will reflect light in a characteristic way. This is known as its luster, and the different varieties of luster are defined by means of comparison with the luster of everyday objects.

Metallic luster is, naturally, like metal. Many of the ore minerals have a metallic luster.

Vitreous luster is like glass. Most of the silicate minerals have a glassy luster.

Resinous luster is similar to plastic.

Pearly luster speaks for itself.

Silky luster is found when a mineral consists of a mass of tiny fibrous crystals that produce a sheen like that of silk.

ANALYZING MINERALS WITH LIGHT

If you slice a rock or a mineral thinly enough it is transparent and, mounted on a glass slide, it can be examined through a special kind of microscope. This technique is the main one used by professional geologists to identify minerals and rocks.

A petrological microscope is basically a conventional desk-mounted optical microscope, with attachments. Below the microscope stage is a polarizing filter – called the polarizer – so only polarized light passes through the specimen. In the barrel of the microscope is a removable polarizing filter – called the analyzer – mounted at right angles to the first, that blocks the polarized light coming up the barrel. As a result, no light is visible by the user. A slice of mineral on the stage, however, may affect the polarization of the light passing through, and it becomes visible to the viewer in a false color – called an interference color. This repolarization of the light by the mineral depends on its crystal symmetry, and the angle at which it lies. The specimen stage of a petrological microscope is a turntable, so that a specimen can be rotated and the effect examined.

Refractive index

This is studied without using the analyzer. Light passing from one medium into another is refracted, or bent, toward the denser medium. The principle is the same as the study of shock waves passing through the Earth after an earthquake (pp. 18–21). The amount by which a ray of light is refracted on entering the mineral is called its refractive index (RI). In a rock slice, where

Right With polarized light passing through, a thin section of the rock schist reveals the individual minerals. Certain kinds of mica show up brown, while large irregular crystals of garnet are a pinkish-gray. Black opaque crystals are tiny fragments of iron ore.

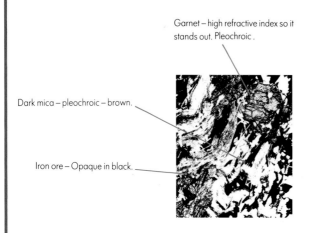

Garnet – high refractive index so it stands out. Pleochroic .

Dark mica – pleochroic – brown.

Iron ore – Opaque in black.

minerals of different refractive indices abut one another, light passing up through the sample is refracted into the mineral with the higher RI. As a result, those minerals with high RIs appear to have a band of brightness around the insides of their margins – the so-called Becke line.

Pleochroism

When a mineral is viewed without using the analyzer, it may show a color. The color may change as the microscope stage is rotated. Such a mineral is said to be pleochroic. Dark mica, for example, will change from dark brown to yellow as it is rotated.

Isotropic minerals

Thin sections of some minerals do not affect the polarity of the light passing through. Polarized light passing through these is blocked by the analyzer, so these minerals show up as dark shapes. Minerals that crystallize in the cubic system show this effect, and are called isotropic.

Anisotropic minerals

Most minerals affect the polarity of the polarized light passing through them, and show up in their interference colors. When that mineral is turned a particular way, the effect on the polarity is minimal and the mineral becomes dark. When the microscope stage is turned to this extinction angle the effect is like a light going out. Some minerals turn dark when they are lined up parallel to the polarization of the light – they show straight extinction. Others go dark when the crystal lies at an angle to the polarization – showing oblique extinction. The studies of these angles are important in identifying individual minerals, but the details are beyond the scope of this book.

Left When a polarizing filter is put in place, affecting the polarized light already passing through the specimen, most of the minerals show up in false colors that can help to identify them. The garnet becomes completely black and the twisted crystals of mica show the strain under which the rock formed. A mass of quartz crystals shows up as a mosaic of gray crystals.

All micas (dark and light) – twisted grain; gaudy interference colors.

Garnet – isotropic is dark.

Quartz – no internal structure. Interference colors – always shades of gray.

Iron ore – still black

COMMON MINERALS

AMPHIBOLE

Silicate of magnesium, iron and calcium, with water in the chemical formula – general formula $(Mg,Fe)_{7-8}(Si_4O_{11})_2(OH)_2$.
Crystal system – Magnesium forms orthorhombic, magnesium and iron forms monoclinic.
Hardness – 5-6
Specific gravity – 3-3.5 increasing with iron content
Luster – Vitreous
Varieties – many different types including hornblende, actinolite, tremolite and asbestos.
Occurrence – Igneous rocks and metamorphic rocks derived from them.

ANHYDRITE

Sulfate of calcium – $CaSO_4$. (Same as gypsum but with no water attached to the $CaSO_4$ molecule.)
Crystal system – Orthorhombic
Hardness – 3-3.5
Specific gravity – 2.93
Luster – Pearly to vitreous
Fracture – Uneven and splintery.
Occurrence – Same as gypsum. (Gypsum may form by alteration of beds of anhydrite.)

APATITE

Phosphate of calcium – $Ca_5F(PO_4)_3$, or $Ca_5Cl(PO_4)_3$
Crystal system – Hexagonal
Hardness – 5
Specific gravity – 3.17-3.23
Luster – Resinous or vitreous
Streak – White
Occurrence – Small amounts in most igneous rocks. A sedimentary rock of pure apatite is mined for fertilizer. Fossil bones are mostly made of apatite.

CALCITE

Carbonate of calcium – $CaCO_3$
Crystal system – Hexagonal
Hardness – 3
Specific gravity – 2.71
Streak – White
Luster – Vitreous to earthy

Fracture – Conchoidal but difficult to observe as it cleaves easily.
Varieties – Iceland spar, a very clear form giving perfect cleavage and showing double refraction which produces a double image of anything seen through it. Dog-tooth spar, sharply pointed crystals. Nail-head spar, with flat-topped crystals. Satin spar, a mass of very fine fibrous crystals.
Occurrence – The main constituent of limestone. Due to its solubility it is often dissolved and redeposited, producing stalactites and stalagmites, hard deposits in kettles, and growths near mineral-rich springs.

CHALCOPYRITE

Sulphide of copper and iron – $CuFeS_2$
Crystal system – Tetragonal

Calcite

Hardness – 3.5-4
Specific gravity – 4.1-4.3
Luster – Metallic
Streak – Greenish black
Fracture – Conchoidal
Occurrence – Usually in veins. The most important copper ore.

CORUNDUM

Oxide of aluminium – Al_2O_3
Crystal system – Hexagonal
Hardness – 9
Specific gravity – 3.9-4.1
Luster – Vitreous or dull.
Fracture – Conchoidal or uneven.
Varieties – Emery, used as an abrasive. Ruby, sapphire, and topaz are discolored by impurities and used as gemstones.
Occurrence – Mostly in thermal metamorphic rocks (see pp. 36–37).

DOLOMITE

Carbonate of calcium and magnesium – $CaCO_3.MgCO_3$.
Crystal system – Hexagonal
Hardness – 3.5-4
Specific gravity – 2.8-2.9
Luster – Crystals are vitreous to pearly, large masses are dull.

Fracture – Conchoidal or uneven.
Varieties – Pearl spar, brown spar and rhomb spar depending on crystal form and color.
Occurrence – As magnesium-rich limestone, like calcite but less soluble.

FELDSPAR

Silicate of aluminum and potassium, sodium, or calcium – $KAlSi_3O_8$ or $NaAlSi_3O_8$ or $CaAl_2Si_2O_8$.
Crystal system – potassium form monoclinic, the rest triclinic. Often in the form of twins.
Hardness – 6-6.5

Twinned crystals of feldspar

Specific gravity – 2.56-2.67
Luster – Pearly or resinous
Varieties – Potassium form called orthoclase or microcline, sodium form called albite, calcium form called anorthite.
Occurrence – Nearly all are found in igneous rocks. Sometimes forming big milky crystals.

FLUORITE

Fluoride of calcium – CaF_2
Crystal system – Cubic
Hardness – 4

Fluorite

Specific gravity – 3-3.25
Luster – Vitreous
Streak – White
Fracture – Conchoidal to uneven.
Varieties – Blue John is a purple decorative variety.
Occurence – Usually deposited in veins cut through the rocks, associated with quartz and some ore minerals. Often shows good cubic crystals.

GALENA

Sulphide of lead – PbS
Crystal system – Cubic
Hardness – 2.5
Specific gravity – 7.4-7.6
Luster – Metallic but often tarnished

Galena

Streak – Lead gray.
Fracture – Flat.
Occurrence – Usually in veins, forming the most important lead ore. Often as good cubic crystals with stepped irregularities on the faces.

Garnet

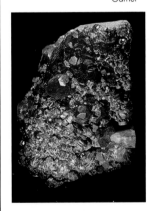

GARNET

Silicate of calcium, magnesium, iron or manganese – General formula $X_3Y_2(SiO_4)_3$, where X is Ca, Mg, Fe or Mn, and Y is Fe, Al, Cr or Ti.
Crystal system – Cubic
Hardness – 6.5-7.5
Specific gravity – 3.5-4.3
Streak – White
Luster – Vitreous

Varieties – Several, including grossular, pyrope, almandine, spessartite, andradite, and uvarovite, depending on the nature of X and Y. Most can be used as an abrasive or as jewelry.
Occurrence – Mostly in metamorphic rocks, where they can often be seen as dark red crystals, sometimes as big as peas.

GYPSUM

Sulphate of calcium – $CaSO_4.2H_2O$
Crystal system – Monoclinic
Hardness – 1.5-2
Specific gravity – 2.3
Luster – Pearly, earthy, or silky.
Varieties – Selenite, well-formed crystals. Alabaster, compact and massive. Satin spar, fibrous and silky.
Occurrence – As chemical sedimentary rock (see pp. 34–35), or in clays.

Gypsum

Hematite

HEMATITE

Oxide of iron – Fe_2O_3

Crystal system – Hexagonal, but rarely seen as good crystals.
Hardness – 5.5-6.5
Specific gravity – 4.9-5.3
Luster – Silky
Streak – Cherry red
Fracture – Uneven
Varieties – Specular iron, black crystals. Kidney ore, fibrous radiating structure like a kidney. Both valuable iron ores.
Occurrence – In pockets in limestones.

HALITE

Rock salt, chloride of sodium – NaCl

Crystal system – Cubic
Hardness – 2-2.5
Specific gravity – 2.2
Luster – Vitreous
Fracture – Conchoidal, brittle, cleaves into perfect cubes.
Occurrence – As a chemical sedimentary rock, deposited as salt-water areas dry out. Sometimes big cubic crystals formed, with stepped concave faces – hopper crystals.

LIMONITE

Oxide of iron – $2Fe_2O_3.3H_2O$ or $Fe_2O_3.nH_2O$. Like hematite but attached to water molecules.

Crystal system – none, forms lumps rather than crystals.
Hardness – 5–5.5
Specific gravity – 3–6.4
Luster – Metallic or dull
Streak – Yellowish brown
Varieties – Bog ore, pea ore, and ocher, the last used as a pigment because of its brown and yellow color.
Occurrence – As a breakdown of other iron ores.

MAGNETITE

Oxide of iron – Fe_2O_4

Crystal system – Cubic
Hardness – 5.5-6.5
Specific gravity – 5.18
Luster – Metallic
Streak – Black
Fracture – Almost conchoidal
Occurrence – Very small crystals in most igneous rocks. Sometimes concentrated in veins, or in sands. In large quantities it is a valuable iron ore. Large lumps act as natural magnets.

MALACHITE

Carbonate of copper – $CuCO_3.Cu(OH)_2$

Crystal system – Monoclinic
Hardness – 3.5-4
Specific gravity – 3.9-4
Luster – Silky or dull

Streak – Pale green
Occurrence – In rounded green masses like bunches of grapes where other copper minerals have weathered.

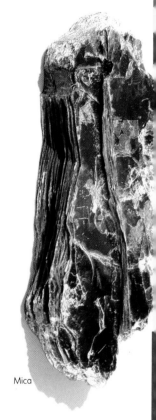

Mica

MICA

Silicate of potassium, aluminium, iron or magnesium – $KAl_2(AlSi_3)O_{10}(OH,F)_2$ or $K(Mg,Fe)_3(AlSi_3)O_{10}(OH,F)_2$.

Crystal system – Monoclinic but looks like hexagonal
Hardness – 2-3
Specific gravity – 2.7-3.1
Luster – Pearly
Varieties – many types (as suggested by the complexity and vagueness of the

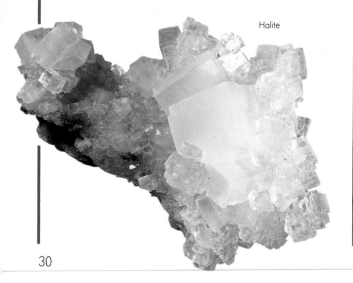

Halite

Pyrite

PYRITES

Iron sulfide – FeS_2

Crystal system – Cubic
Hardness – 6-5.6
Specific gravity – 4.8-5.1
Luster – Metallic and shiny
Streak – Greenish or brownish-black
Fracture – Conchoidal
Occurrence – Can form good cubic crystals with corrugated faces. Often in veins, where its brassy appearance can be mistaken for gold, giving it the common name fool's gold.

PYROXENE

Silicate of magnesium, iron and calcium – $(Mg,Fe,Ca)_2(Al,Si)_2O_6$. Very similar to amphibole but without the water part in the chemical formula.
Properties and occurrence very similar to amphibole.
Varieties – Enstatite, augite and diopside.

QUARTZ

Silica – oxide of silicon. SiO_2
Crystal system – Hexagonal. Good crystals have hexagonal prisms with hexagonal pyramids at the ends.
Hardness – 7
Specific gravity – 2.65
Luster – Vitreous
Fracture – conchoidal
Varieties – In its pure form, it is clear and called rock crystal. Turned opaque by the inclusion of tiny air bubbles, it becomes milky quartz. Mineral impurities turn it brown – smoky quartz or Cairngorm – or violet – amethyst.

Occurence – Common in some igneous rocks – the so-called acidic rocks (see p. 32). Often forms the grains of sand, and of sandstone. Commonest substance of veins or infillings. Good crystals only found in cavities.

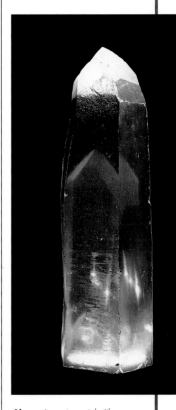

Above A quartz crystal with an internal ghost crystal.

chemical formula).
Mainly white mica muscovite with potassium, and black mica biotite with iron.
Occurrence – Igneous and metamorphic rocks. Atoms arranged in sheets, and so it flakes easily (strong cleavage). Rock containing mica often glistens in the sun.

OLIVINE

Silicate of magnesium or iron, or a mixture of the two – Mg_2SiO_4 or Fe_2SiO_4, usually expressed as $(Mg,Fe)_2SiO_4$.

Crystal system – orthorhombic
Hardness – 6-7
Specific gravity – 3.2-4.3
Luster – Vitreous
Streak – clear
Fracture – conchoidal
Varieties – Magnesium form called forsterite, iron form called fayalite.
Occurence – Only in basic igneous rocks (see pp. 32–33). Often as crystals that are cracked, altered to the mineral serpentine $(Mg_6Si_4O_{10}(OH)_8)$ along the cracks.

THE ROCK CYCLE AND IGNEOUS ROCKS

The surface of the Earth is continually being created and destroyed – not usually in any way that can be perceived by human observation, but over millions, tens of millions, and hundreds of millions of years. The surface is constantly being built up of masses of mineral material, and these mineral masses are just as constantly crumbling away. The masses of mineral material are called the rocks.

We recognize three different types of rocks, depending on how they are formed.

Igneous rocks

Molten material welling up from within the Earth may eventually cool and solidify. The solid mass that results is called an igneous rock.

There are two main types of igneous rock – intrusive and extrusive.

Intrusive igneous rocks form as the molten material pushes its way upward through the rocks, cutting across or squeezing between them, and solidifying before reaching the surface. If such rocks cool slowly, they will be coarse and have mineral crystals big enough to be seen with the naked eye. If they cool quickly, they will be fine-grained. Sometimes the molten mass begins to cool slowly, crystals of one mineral begin to form, and then the whole lot is thrust into another area where it cools quickly. This gives a porphyritic texture, with big crystals in a fine groundmass.

Acidic	A rock that contains more than 66 per cent silica.
Intermediate	One that contains 66 to 52 per cent silica.
Basic	One containing 52 to 45 per cent silica.
Ultrabasic	A rock that contains less than 45 per cent silica.

Extrusive rocks, on the other hand, are those that form when the liquid erupts at the surface, as from a volcano. These are always much finer than the intrusive forms. All lavas are extrusive igneous rocks.

Another way of classifying igneous rocks is by their chemical composition. They can be rich in

Right An extrusive igneous rock forms as molten material erupts from beneath the Earth's surface and solidifies, as here in the volcano Stromboli in the Mediterranean. **Left** An intrusive igneous rock solidifies underground, and we do not see it at the surface until the overlying rocks have been worn away. Here a mass of andesite protrudes from the surrounding softer rocks in Wyoming.

silica, or poor in silica. (Those that are poor in silica still have a very high proportion of silica in them, but not as high as the others.) Such a classification involves a somewhat misleading nomenclature which convention dictates that we must use.

The terms derive from an old chemical idea that rocks are the salts of some kind of "silicious acid" – total nonsense by modern understanding, but it does leave us with a useful and workable classification.

By applying both of the above classifications, and combining the grain of the rock with its chemical composition, we can start to define the most common types of igneous rock.

Intrusive acidic rock shows large crystals, many of which are quartz. Acidic rocks tend to be lightly colored, because of the presence of quartz. Basic and ultrabasic rocks are dark. There are no ultrabasic extrusive rocks – ultrabasic rocks are rare at the Earth's surface but are thought to be the main constituent of the mantle.

	Extrusive	Intrusive
Acidic	Rhyolite	Granite
Intermediate	Andesite	Diorite
Basic	Basalt	Gabbro
Ultrabasic		Peridotite

As a rule acidic and intermediate rocks form by the solidification of molten crustal material. Basic rocks are more likely to form from molten material brought up from the mantle.

This is a simplification. The true situation is much more complex. A solidifying melt goes through many stages of a process called fractionation before it becomes an igneous rock. As the mass cools the first minerals to crystallize out are usually those relatively low in silica, such as olivine, pyroxene, and amphibole. These can then sink to the bottom of the mass leaving a liquid that has become relatively rich in silica, and this may erupt toward the surface and form acidic rocks. A liquid rising through the tubes of a volcano will find the surrounding pressure decreasing. The gassy components will fizz off, like the bubbles in champagne when the cork is popped. A volcanic eruption will be accompanied by great blasts of gas and steam, and will produce an extrusive igneous rock that has little chemical similarity to the melt that spawned it.

Something to do

Dissolve some alum, or copper sulfate in warm water to produce a concentrated solution, as you did in the crystallization experiment on pp. 22–23. Now let it cool as slowly as possible. You should see large crystals forming. Now do the same and let the solution cool quickly. The crystals should now be so small as to be indistinguishable to the naked eye. This demonstrates the difference between the grain of an intrusive and an extrusive igneous rock.

THE ROCK CYCLE AND SEDIMENTARY ROCKS

It is hard being a rocky outcrop. An exposed rock face is continually, day in, day out, bombarded by the worst that the weather can throw at it. Water rains down and soaks into the pores, frost freezes it, cracking the pores open, acid dissolved in the rainwater reacts with certain of the minerals, the heat of the sun expands and contracts the surface, falling rocks from above chip pieces away. Sooner or later any exposed rock will turn to rubble or dust.

But that is not the end of the story. The rubble and dust are carried away by rivers or by wind, and eventually deposited somewhere else, where they will all eventually be consolidated and turned into rock. This is the genesis of the second type of rock – the sedimentary rock.

As with the igneous rocks, there are various kinds of sedimentary rocks. Three main types are recognized.

First there are the clastic sedimentary rocks. These result from the processes outlined above. They are the result of some preexisting rock's demise, and the recementing of its fragments to form a new rock. There are many kinds of clastic sedimentary rock, depending mostly on the size of

Above Conglomerate is a coarse-grained clastic sedimentary rock, made up of rubble or gravel buried, compacted, and cemented into a whole.

Below Coal is an example of a biogenic sedimentary rock. It is made up of the accumulation of fragments of plants that were once alive.

Particle diameter
greater than 0.02 in (2 mm)
These can be boulders, cobbles or pebbles. The resulting sedimentary rock is conglomerate if the particles are rounded, or breccia if they are jagged and uneven.
0.00015 – 0.02 in (1/256 mm – 2 mm)
These are the sands, and produce the sandstones of various types.
smaller than 0.00015 in (1/256mm)
Muds and clays come into this category, producing the clastic sedimentary rock called shale if it is well-bedded, mudstone if it is flaky, and clay if it has no structure.

the fragments that are cemented together.

The second kind of sedimentary rock is biogenic sedimentary rock. Such a rock is built up of material produced by living organisms. Coal is one of the most familiar, consisting of carbon derived from masses of ancient vegetable matter. Certain limestones, when examined, are seen to consist almost entirely of fragments of fossil shells, or even of coral material cemented together as a reef.

Above Sandstone is a medium-grained clastic rock formed from sand on a beach, in a river bed or, in this case, a desert.

Below Oolitic limestone is a chemical sedimentary rock, formed from calcite deposited on particles on an ancient seabed.

It is a considerable step between a layer of rocky fragments and a sedimentary rock – between a bed of sand and a sandstone, between a heap of seashells and a shelly limestone. That step is called lithification and can be achieved by a number of processes. First the bed of fragments is buried by subsequent beds, and is then compressed by the weight of the new sediments on top of it. This has the effect of squeezing the fragments into one another and squeezing out the air in between. Groundwater, seeping through the mass, may be rich in dissolved minerals such as calcite, and these can be deposited as tiny crystals in what spaces are left, forming a natural cement that binds the fragments together. Alternatively the groundwater may act on the fragments themselves, and induce the mineral crystals in them to continue to grow until they all become a single interlocked solid mass. This sometimes happens with quartz grains in sandstone.

In any case, sedimentary rocks are usually very recognizable in the field, because they lie in distinct layers, or beds. The analysis of the nature of the beds can tell us much about the surface of the Earth in times gone past (see pp. 50–69).

The last class is the chemical sedimentary rock. This is produced by inorganic chemical material being deposited on the floor of a sea or a lake and building up into a solid mass. Rock salt and anhydrite occur when dissolved salts in a body of water are deposited as the water evaporates away. Certain limestones consist of calcite that is deposited in shallow water when currents bring calcite-rich waters into areas where the water chemistry is different.

Something to do

Make equal weights of sand and dry plaster of Paris. Pour it into a transparent plastic container to make a layer of about one inch (2 cm). Now add a mixture of fine gravel and plaster. Do the same for various mixtures of colored sand, or pebbles, all with plaster of Paris. Throw in a seashell or two on a layer.

Now fill the container with water. Next day you will have your own sedimentary rocks. The water will have caused the plaster to recrystallize and cement the whole mass together. The seashells will be present as fossils.

THE ROCK CYCLE AND METAMORPHIC ROCKS

The third category is rock that has been changed. The minerals in an igneous rock, or the particles in a sedimentary rock, may look fairly permanent, but in certain circumstances, for example, under extreme conditions of pressure or temperature, they can change and recrystallize into something new. The new rock produced is called a metamorphic rock.

There are two types of metamorphic rock. The first is regional metamorphic rock in which the altering force is one of pressure rather than of temperature. These are found deep within the interior of mountain chains, and are believed to constitute the lower parts of the crust. Vast tracts of the Earth's crust have been altered in this way, hence the name. Different degrees of pressure produce different grades of metamorphic rock. Slight pressure – and the word "slight" is a comparative term – will produce a low-grade metamorphic rock, in which the only difference will be that the minerals will have been realigned in a different direction. Often this produces flat crystals of mica that are orientated according to the direction of the applied pressure. The result is a rock that has planes of weakness running in one direction, and which can split easily into flat slabs. Slate and phyllite (see pp. 112-113) are typical low-grade metamorphic rocks. At the other end of the scale, intense pressure will completely change the mineralogical makeup of the rock and produce a high-grade metamorphic rock. The chemical components may recrystallize into a totally different set of minerals from the original rock and the new minerals may form in distinct bands, often crumpled and contorted as evidence of the great pressures involved. Gneiss (see pp. 116-117) is the typical high-grade metamorphic rock showing distinct banding.

A typical sequence of rocks – from unconsolidated sediment, through sedimentary rock, through different grades of metamorphic rock, depending on the depth in the crust at which different conditions are found – is shown opposite.

The last rock in this series, hornfels, in fact belongs to the second type of metamorphic rock – the thermal metamorphic rock, sometimes called the contact metamorphic rock. Heat is the most important influence in the formation of such rocks. As a result, thermal metamorphic rocks are less common and much more restricted in distribution than their regional counterparts. The usual place to find them is at the edge of an intrusive igneous rock, where the heat of the cooling mass has cooked the native rocks at each side. This will produce a metamorphic auriole around the igneous rock, which may only be an inch or two (a few centimeters) wide. Unlike regional metamorphic rock, thermal metamorphic rock shows no internal structure, and can often be mistaken for an igneous rock.

Different minerals crystallize at different temperatures in a metamorphic auriole, and so the mineralogy of the rock close to the intrusion will be different from that farther away. The amount of heat given off as the body cools is another important variable. The chemical constituent of the original rock determines the new minerals that are formed. In a sandstone that contains nothing but quartz fragments, the quartz recrystallizes in a more compact mosaic, forming the thermal metamorphic rock called quartzite. In a pure limestone the calcite will recrystallize to form marble. Displacement or dynamic metamorphism is local alteration caused by friction as

Surface
Mud
3 miles (5 km) deep
Shale (sedimentary)
6 miles (10 km) deep
Slate (low grade metamorphic). Different kinds of micas develop.
9 miles (15 km) deep
Schist. Garnet appears.
12 miles (20 km) deep
Gneiss (high-grade metamorphic). Staurolite forms.
15½ miles (25 km) deep
Hornfels. Strange rare minerals such as cordierite appear.

Two examples of regional metamorphic rock. The schist **right** shows well the new minerals, such as the red staurolite and the pale blue kyanite, produced by the intense metamorphic action. The gneiss **below** shows how the pressure under which the rock formed has produced new minerals and then pulled them out into elongated structures.

one mass slides over another.

Then things become complicated. When a metamorphic rock is itself metamorphosed, the process is called polymetamorphism.

In all this compexity the important point to note is that metamorphism takes place in solid rock. The minerals recrystallize without passing through a molten phase. Should the minerals, at any stage of the operation, melt, then the result would not be a metamorphic rock, but an igneous one.

Something to do

Make a snowball. Pick up a handful of snow and compress it to form a lump sturdy enough to be used as a missile. In the center of the snowball, the light fluffy crystals of snow will have recrystallized in a more compact form, because of the pressure of your hands. This is quite close to the metamorphic process.

EARTH MOVEMENTS

The Earth is not static. The forces that shape its surface are at work all the time, slowly heaving up the mountains and just as slowly wearing them away.

PLATE TECTONICS

If you boil a saucepan of soup, froth and scum forms on the surface, and it moves about in lumps in response to the convection currents churning about below. This is a little like what has been going on at the surface of the Earth.

The Earth's crust is constantly being destroyed and renewed – not just the rocks of the continents, but the entire outer covering of the globe. Imagine the surface of the Earth as consisting of a number of panels, or plates, like the panels of a football. Imagine molten material welling up along one seam of a panel, and then solidifying to form the material of the panel itself. Imagine then that the newly formed panel is constantly moving away from that seam, and is buckling down beneath the next panel and being destroyed. That is what is happening to the Earth's surface.

All this activity takes place on the floor of the oceans, and we only found out about this back in the 1960s. Throughout the oceans there is a system of ridges. These are the places where the new surface material is being created. In other areas,

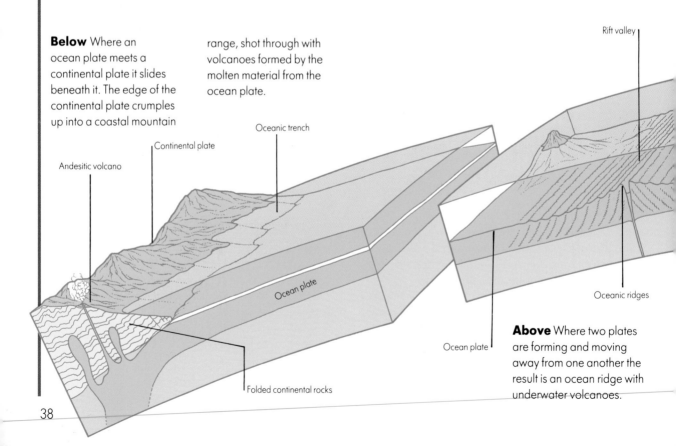

Below Where an ocean plate meets a continental plate it slides beneath it. The edge of the continental plate crumples up into a coastal mountain range, shot through with volcanoes formed by the molten material from the ocean plate.

Andesitic volcano

Continental plate

Oceanic trench

Ocean plate

Folded continental rocks

Rift valley

Oceanic ridges

Ocean plate

Above Where two plates are forming and moving away from one another the result is an ocean ridge with underwater volcanoes.

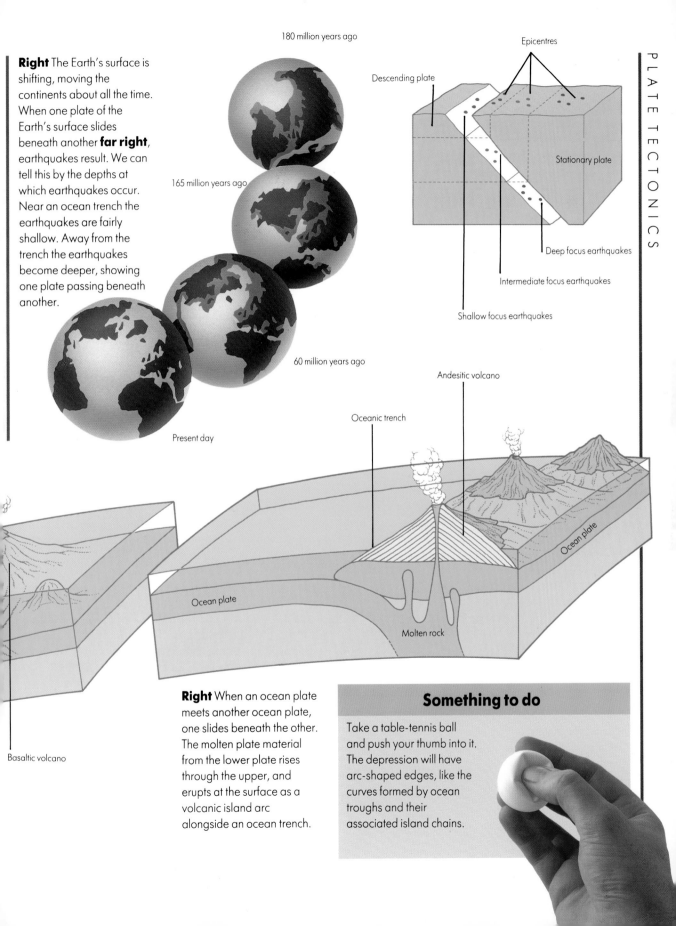

Right The Earth's surface is shifting, moving the continents about all the time. When one plate of the Earth's surface slides beneath another **far right**, earthquakes result. We can tell this by the depths at which earthquakes occur. Near an ocean trench the earthquakes are fairly shallow. Away from the trench the earthquakes become deeper, showing one plate passing beneath another.

180 million years ago

165 million years ago

60 million years ago

Present day

Epicentres

Descending plate

Stationary plate

Deep focus earthquakes

Intermediate focus earthquakes

Shallow focus earthquakes

Andesitic volcano

Oceanic trench

Ocean plate

Ocean plate

Molten rock

Basaltic volcano

Right When an ocean plate meets another ocean plate, one slides beneath the other. The molten plate material from the lower plate rises through the upper, and erupts at the surface as a volcanic island arc alongside an ocean trench.

Something to do

Take a table-tennis ball and push your thumb into it. The depression will have arc-shaped edges, like the curves formed by ocean troughs and their associated island chains.

Below The trace of an ocean ridge is seldom straight, yet it does not sweep around in great smooth curves. Instead it forms a step-like zigzag pattern, with small sections of ridge offset from one another and connected by "transform faults." This is caused by the geometry involved as the plates expand in each direction from the ridge.

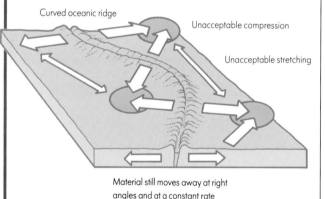

Straight oceanic ridge

Material moves away from ridge at right angles and at a constant rate

Curved oceanic ridge

Unacceptable compression

Unacceptable stretching

Material still moves away at right angles and at a constant rate

The stresses resolve themselves

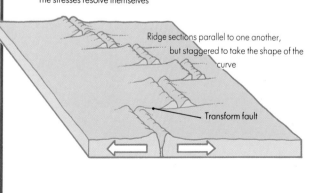

Ridge sections parallel to one another, but staggered to take the shape of the curve

Transform fault

notably around the edges of the Pacific, there are deep troughs. These are the places where the surface material is being drawn down and destroyed. The continents are sitting embedded in these moving plates and are being shuffled around by the movement all the time.

The material of these plates consists of the crust and the topmost solid layer of the mantle which together are called the lithosphere. They move about on a spongy region of the mantle (see pp. 14–15) which we call the asthenosphere.

Where the material is being created, and where it is being destroyed, there is volcanic and earthquake activity. The newly formed plate material makes up the ocean floor, and so nowhere on the Earth's surface is the ocean floor more than about 200 million years old. The younger areas are found closest to the ocean ridges. Where one ocean plate is destroyed beneath another, molten plate material rises and forms chains of volcanic islands like those that festoon the edge of the Pacific. These form in arc-shapes because of the three-dimensional geometry involved in curling the broken edge of a sphere downward.

The continental crust – the sial – is less dense than the ocean crust – the sima – and so it tends to float in it. It is too light to be drawn downward into the mantle – just as a piece of toast floating in a sink will not be drawn down the vortex of the plug hole – and so when the continent is carried to an ocean trench it stays there, crumpling up with the continuing movement. This is the reason for the active mountain chains, perforated by volcanoes, that we find at the edges of some continents. When two continents are brought together they fuse into one supercontinent, uplifting a massive mountain range along the joint. The Himalayas are an example of such a range that is now being formed. The Urals represent one that was formed 300 million years ago and is now being worn away.

The theory that oceanic crust is continuously being created and destroyed, and continents shift about sounds plausible, but what is the evidence?

Some inkling of the idea had been there for centuries. The English statesman and natural philosopher Francis Bacon noted the similarity between the west coast of Africa and the east coast of South America, and pointed it out in 1620. Further observations in subsequent years led to the postulation of continental drift by the German meteorologist Alfred Wegener in 1912. He sug-

Offset ridges and transform faults

The new material produced at an ocean ridge spreads away at right angles to the ridge. If it is curved, the ridge breaks into a series of steps, and the new material formed at each step can move away in the same direction in every section.

You can simulate the movement in this paper model. Enlarge both diagrams to twice the size shown. Cut slots in the baseboard to represent the staggered zone of upwelling. Now cut and fold the bottom diagram as indicated. This represents the new ocean crust material. Insert the folds into the slots as shown. Now you can pull the sides apart and as the paper emerges from the slots it simulates the movement of the new ocean material. The different slices slide past one another in a structure called a transform fault.

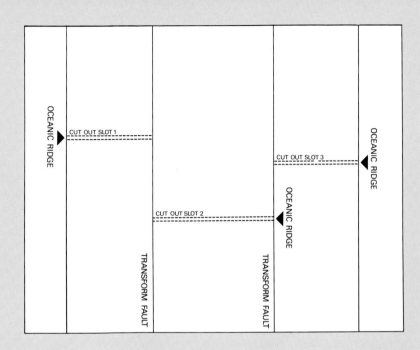

OCEANIC RIDGE

CUT OUT SLOT 1

OCEANIC RIDGE

CUT OUT SLOT 3

CUT OUT SLOT 2

OCEANIC RIDGE

TRANSFORM FAULT

TRANSFORM FAULT

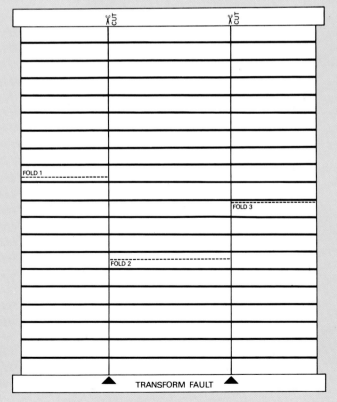

CUT

CUT

FOLD 1

FOLD 3

FOLD 2

TRANSFORM FAULT

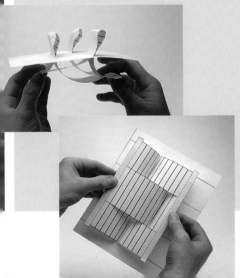

gested that all the continents had once been a single supercontinent but had since split up and the fragments drifted to their present positions. He was, however, at a loss to account for the mechanism. By the middle of the 20th century proof was there in abundance – mainly the continuation of geological structures from one continent to another distant continent and the same fossils found on widely separated continents.

Then, in the 1960s, the British geologists Fred Vine and Drummond Matthews were studying the ocean floors. They found that the rocks on each side of the ocean ridges became older farther away from the ridge crest. Many times in the past the Earth's magnetic field has reversed, north pole for south. Magnetic particles in the rocks formed at a particular time will show the contemporary polarity of the field. Vine and Matthews found the ocean floor to be divided into stripes, each stripe magnetized in a different direction. The pattern of magnetization at one side of the ridge was the mirror image of that at the other. They recognized that this was due to the fact that the ocean floors were growing continuously from their ridge crests, and called the phenomenon seafloor spreading.

It was then but a short step to combining the old concept of continental drift with the newer one of seafloor spreading to produce a comprehensive theory of plate tectonics.

Something to do

Reports of earthquakes come through regularly. Keep a note of them by the age-old method of sticking little flags on a map of the world. If your map is one that shows the plate boundaries you will soon find that the earthquake distribution follows the pattern of the edges of the plates.

Further demonstration has been made regularly. Earthquakes along the west coast of South America and Japan are very shallow in the region of the local ocean trench, but become deeper farther under the land. The greatest depths are about 435 miles (700 km). This indicates the position of the descending lithospheric plate as it is being destroyed. The site where such an event is presently occurring is called the subduction zone or the Benioff zone. Submarine photographs reveal that rocks near the ocean ridges are clean and new, but have a deeper cover of sediment farther away from the ridge. At about 6 miles (10 km) from the ridge crest they are totally blanketed by sediment accumulated over thousands of years.

The worldwide distribution of earthquake centers and of volcanic eruptions is remarkably close to the pattern of constructive and destructive plate boundaries. Such events are witnesses to the great forces that are tearing at the surface of our planet as new material is being made and older material destroyed.

Very careful analysis can show the speed of tectonic movements. The fastest is the movement of the Australian plate, which is shifting the continent of Australia northward at a rate of 6½ in (17 cm) per year. The movement of the Atlantic, at ½–1 in (1–2 cm) a year on each side, is more typical. The Atlantic Ocean is now 32½ ft (10 m) wider than it was in Columbus's time!

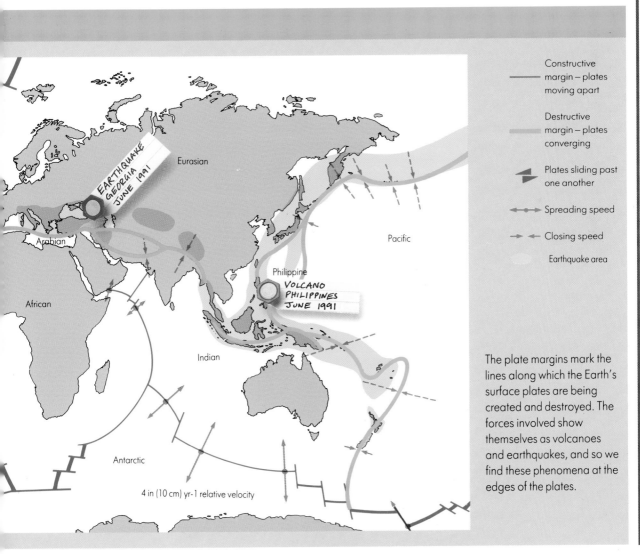

The plate margins mark the lines along which the Earth's surface plates are being created and destroyed. The forces involved show themselves as volcanoes and earthquakes, and so we find these phenomena at the edges of the plates.

PLATE TECTONICS

43

FIELD EQUIPMENT

The field geologist should be physically fit, and be able to cope with outdoor conditions. Camping should be second nature. It would be useful to have several of the outdoor skills, such as rock-climbing and mountaineering, and to be able to handle a four-wheel-drive vehicle. It is also useful to know how to ride a horse.

This all sounds as if the practical geologist should be a marine commando, or at the very least a boy scout. These backgrounds would as a matter of fact be useful for field work in the Askja Caldera in central Iceland, or on the face of the Great Rift Valley in Tanzania. However, much valuable field work can be done in accessible localities, for which you will need a minimum of preparation.

Food and clothes

Such preparation includes suitable clothes. Dress must suit locale – waterproof parkas and pants and warm underwear for cold wet regions, and light reflective clothes and sun hats for hot sunny areas. Make them old clothes – it is a dirty business. Good strong hiking boots are a must, because you will inevitably be walking over rough, rocky terrain. Much of this is common sense, applicable to any outdoor work or activity. An addition to any wardrobe would be a pair of tough gloves. Your bare hands will suffer after handling rocks all day.

Food must also be considered. It may be that your work will be done not far from some eating place, in which case there is no problem. Otherwise plan sufficient amounts of food and drink, and allow for these when packing your equipment.

Basic equipment

The equipment should be kept to a minimum, especially if you are considering bringing back specimens. A backpack full of rocks will be heavy.

Geological hammers are essential. Indiscriminate bashing at a rock-face is the sign of the amateur. Equally, you should not destroy natural habitats without good reason; do not unnecessarily destroy plants (even lichens are of scientific importance). However, you must break open a rock in order to see its true nature: any rock that has been left will be weathered on the surface and probably covered in moss or lichen. Only a fresh face will show what the rock is really like. Take a penknife as well, for separating laminae of shale or crystals of mica.

The practical geologist should wear outdoor clothes suitable for the locale and conditions. Wet and cold weather calls for waterproof garments, as on Mount Saint Helens **right**. Hot conditions, such as those in the Philippines **far right**, demand lighter clothes. In all cases good heavy boots are essential.

1 The square end of a geological hammer is for smashing open massive rocks and for tapping chisels, and the other is for splitting sedimentary beds. Perhaps the most versatile size is 2.2 lb (1 kg). **2** A sheath is handy for carrying the hammer. **3** Cold chisels, useful for more delicate splitting work. **4** Geological pins are useful for removing individual crystals or tiny fossils.

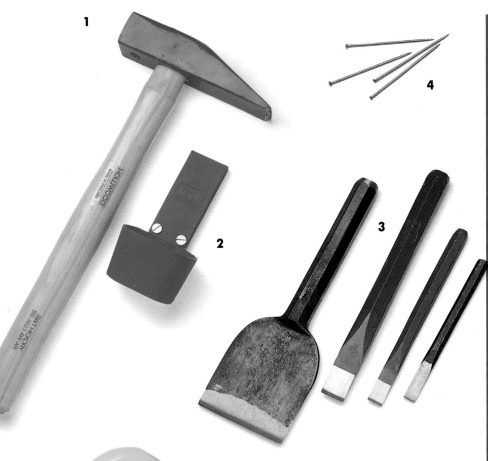

Safety tips

Safety tips

Do not try to cut corners by using a carpenter's hammer. The steel is too soft and will distort and split in no time at all. Never use the head of one geological hammer as a chisel, while hitting it with another. The hammers are not tempered during manufacture to take that sort of treatment. Accidents have been caused by flying steel shards from hammer heads splitting under such misuse.

When smashing up rocks it is important to wear a pair of goggles, to keep splinters out of your eyes. Wear a safety helmet when you are working at the bottom of a cliff face.

Although it is best to travel in groups, take a whistle, the sound of which will carry farther than your voice. Take first-aid supplies also.

USING FIELD EQUIPMENT

Smaller pieces of equipment have specialized uses.

Measuring dip

An important concept in the study of geology is that of dip. Dip is defined as the angle at which a stratum is inclined from the horizontal. The direction of dip on an exposed surface can be found by watching which way water runs down it. The angle of dip is measured by a clinometer. Such an instrument usually comprises a straight edge that is lined up against the dip of the rock, and a weight that shows the vertical. The angle of dip can be read off. Commercially produced clinometers are often combined with compasses.

Measuring strike

A compass is a valuable aid in field geology. It is, of course, used to orientate your maps. It is also used to note the strike of the outcrops. The strike can be thought of as the reciprocal of the dip. It is the line that a dipping bed makes with the horizontal – the waterline that would be formed if the bed dipped into a lake. The strike is very useful in geological mapping (pp. 122–133).

Observing and measuring

No field kit would be complete without a hand lens. The most useful magnifications are 8× or 10×, and the most convenient type is the one that folds away into a protective slot on the handle. When you find a mineral crystal or a tiny fossil worth looking at, you can whip out the lens very easily. Hold the lens close to your eye and focus it by moving the specimen toward and away from it.

Tape measures of one sort or another are essential. A very long surveyor's tape is only useful if you are doing detailed mapping and surveying. An engineer's steel tape is probably the most versatile, but a smaller angler's tape may be all you need and would be much lighter to carry.

Recording

You must record your finds. A hardcover notebook is essential. This should not be too big or it will become unwieldy – something about 4 in × 8 in (10 cm × 20 cm) is ideal. In this you can write your observations and make your sketches. Sketches may be made more conveniently on separate paper

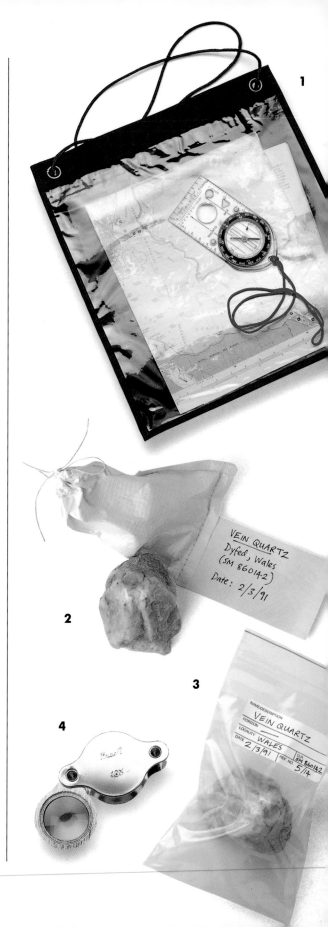

clipped to a clipboard. This can double as a map case for the maps that you will be using. Needless to say, you will need a plentiful supply of pens and pencils. Another piece of recording equipment is the camera, complete with suitable lenses and film for the type of work you are anticipating. So-called "point-and-shoot" cameras are increasingly used by professional geologists because of their simplicity and dust-protective design.

Finally, you must take wrapping materials for your specimens. Loose rocks in a bag will rattle about and become chipped and worn, ending up covered in white streaks of ground powder. Separate little linen bags are used by professional geologists, but newspaper does just as well.

The bag that carries all this must be big enough and sturdy enough to carry your equipment and your specimens as well. A canvas haversack is best. A nylon one will do, but it will wear out more quickly as the heavy items pull it into holes.

Make your own clinometer

A simple clinometer can be made by gluing a protractor to a piece of cardboard as shown. You must ensure that the horizontal of the protractor is at exactly right angles to a marked straight edge of the cardboard. Drill a hole at the protractor's center and thread a weighted line through it.

To use the clinometer, simply place the straight edge on the rock surface, ensuring that it is parallel to the dip, and read off the angle of dip on the protractor scale.

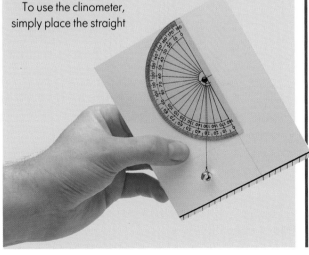

Field equipment should be kept to a minimum, but here are a few essentials. **1**. The map case should be waterproof and a handy size. The most useful compass is the kind used in orienteering. **2** and **3**. Individual labeled bags, or sheets of newspaper should be used for carrying specimens. **4**. A small hand lens is useful for on-the-spot identifications. **5**. Measuring and recording equipment should be easily portable. **6**. The bag for all this must be robust and easy to carry.

PREPARING A FIELD NOTEBOOK

Your field notebook must be sturdy and have a hard, and preferably water-resistant, cover. There is nothing worse than standing in the rain trying to fill in information on soggy pages that are falling out of their binding – and watching the ink run because you brought the wrong kind of pen as well as the wrong kind of notebook.

It is not a good idea to take rough notes in the field and write them up afterward. Rough notes tend to be ambiguous, and their actual meaning becomes lost if you try to reconstruct your observations later. Write down as much as possible while you are on site and while the rocks and structures are there in front of you.

The entries in your field notebook must be comprehensible to you if you want to go back to the site a few years later, or if you give the notebook to another geologist who wants to follow up your work in the same area.

First you must record the location of your site. In the United States this means using a U.S.G.S. (United States Geological Survey) quadrangle map.

Make plenty of illustrations in your field notebook. One picture is worth a thousand words of description. Besides pictures of outcrops, sketch-maps of the sites are also useful.

More detailed sketches can be made on separate large sheets of paper, attached to your map case, and more information can be put on your field map (see "Mapping").

All these can be put together to produce a comprehensive picture of the geology of an area, and can all be referred to when you write your final report.

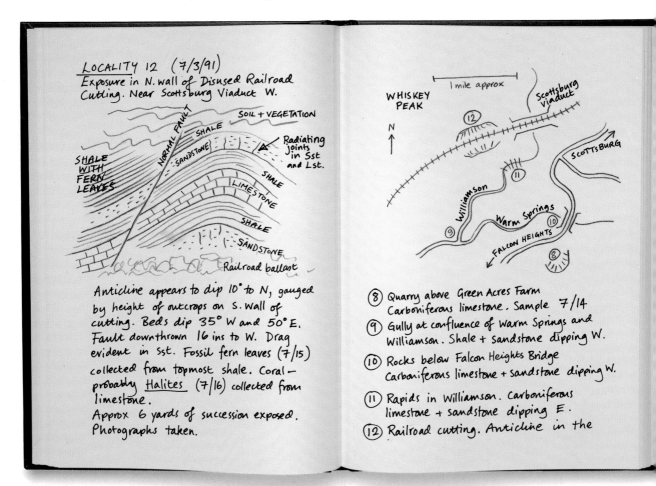

LOCALITY 12 (7/3/91)
Exposure in N. wall of Disused Railroad Cutting. Near Scottsburg Viaduct W.

SOIL + VEGETATION
NORMAL FAULT
SHALE
SANDSTONE
Radiating joints in Sst and Lst.
SHALE WITH FERN LEAVES
SHALE
LIMESTONE
SHALE
SANDSTONE
Railroad ballast

Anticline appears to dip 10° to N, gauged by height of outcrops on S. wall of cutting. Beds dip 35° W and 50° E. Fault downthrown 16 ins to W. Drag evident in Sst. Fossil fern leaves (7/15) collected from topmost shale. Coral – probably Halites (7/16) collected from limestone.
Approx 6 yards of succession exposed. Photographs taken.

WHISKEY PEAK
N
1 mile approx
Scottsburg Viaduct
12
Scottsburg
11
Williamson
Warm Springs
10
Falcon Heights
9
8

⑧ Quarry above Green Acres Farm Carboniferous limestone. Sample 7/14

⑨ Gully at confluence of Warm Springs and Williamson. Shale + sandstone dipping W.

⑩ Rocks below Falcon Heights Bridge Carboniferous limestone + sandstone dipping W.

⑪ Rapids in Williamson. Carboniferous limestone + sandstone dipping E.

⑫ Railroad cutting. Anticline in the

Where am I?

Most maps use a grid reference system. This is composed of horizontal lines and vertical lines. Each line is given a two figure number. By quoting the two figures on the horizontal axis, and the two figures on the vertical axis, you can identify a particular square on the map. Thus grid reference 3479 will define a square bounded on two sides by the vertical line 34 and the horizontal line 79. (It is conventional always to take the number of the vertical line first, as read off the bottom margin of the map – and then the horizontal line as read off the left-hand margin.) Depending on the scale of the map, this square can represent a relatively small to relatively large area on the ground. To be more precise you must, in your mind's eye, divide each axis of that square into tenths and define the point in the square by another grid reference that you add to your four-figure reference. Thus, the six-figure reference 345793 refers to a point five tenths of the way across the square from vertical line 34, and three tenths of the way up from horizontal line 79.

Left and right The field notebook must contain all your observations made on the spot, with maps, sketches, and diagrams of whatever seems important at the time. Remember that it is a document that you will want to refer to later, and so you must include all the relevant information – in a form that is readable to yourself and to anyone else who may wish to work from it. File index cards should carry a catalog number, a brief description of what the specimen is, where it was found, and the date on which it was collected.

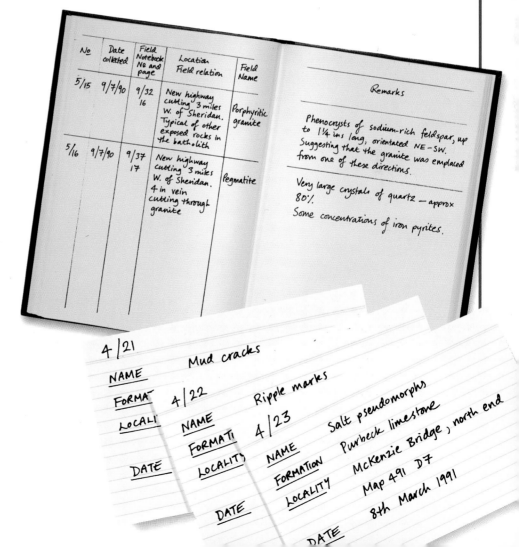

ROCK DEFORMATIONS

The movement of the continents has left its mark on their constituent rocks. Continents that have crumpled up at the edge of a subduction zone, or have ground past one another, or have crushed up against one another to form a single landmass, contain rocks that have been affected in various ways. In the extreme form, such deformation causes metamorphism (pp. 36–37), but most often rock units are folded or simply inclined.

Folded rocks

When a layered sequence of rocks is compressed, it fractures or folds. Something as apparently solid and brittle as a bed of rock can actually fold. Push a tablecloth across the table and it will develop parallel wrinkles, at right angles to the direction of push. This is similar to what happens to the rocks of the continents. In fact, extreme forms of folds, where the different waves of rock collapse over

one another as deep within the Alps' center core, are called nappes (from the French for tablecloth). A few general terms are essential here. When a fold sags downward it is called a syncline. When it arches upward it is called an anticline. See the box below for further technical explanation.

We do not usually find an isolated syncline or anticline. More often they are found in a series, one following the other. The fold can be symmetrical if each flank dips at the same angle, asymmetrical if one flank is steeper than the other, or recumbent if it has turned over upon itself. Isoclinal folds are those that are so compressed that the limbs are parallel to one another.

Often the fold occurs in three dimensions, forming either a basin or a dome.

Usually we do not see a fold in its entirety in the field. We can deduce its presence by recognizing the same beds dipping in different directions not far from one another.

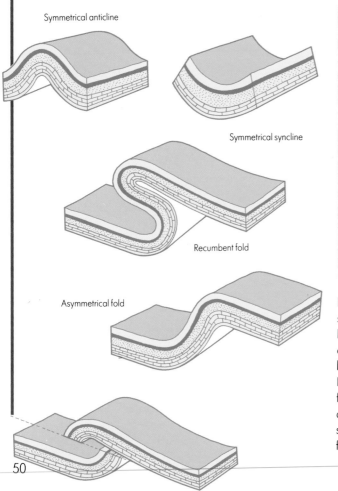

Symmetrical anticline

Symmetrical syncline

Recumbent fold

Asymmetrical fold

Left The simplest folds are symmetrical, each side being a mirror image of the other, as the anticline **top left** and syncline **top right** Intense sideways pressure forms, in turn, asymmetric and recumbent folds, finally shearing off the limb of the fold as a thrust.

Above In nature, folds are rarely simple. It is unusual to find the textbook syncline and anticline in isolation. In this example, pressure from left to right has twisted the strata into a recumbent fold at this locality in the Cape Fold Belt in South Africa.

Things to look for in a fold

Below In the field there are a number of features that we can examine in a fold. Most of them can help us to understand the forces that produced the fold in the first place. The orientation of the fold can tell us the direction from which the forces came, and the cracks and joints can show the stresses to which the rocks were subjected.

Axis The official definition of the axis is the line that moves parallel to itself to generate the fold. In more practical terms, it is the line around which the fold is bent.

Plunge If the axis is not horizontal, it makes an angle to the horizontal called the plunge.

Axial plane This is the plane joining up the axes of the various beds in the fold. It may be vertical in a symmetrical fold, or inclined in an asymmetrical fold.

Competent beds Those that tend to hold their original shape when deformed. They tend to break rather than bend.

Incompetent beds Those that deform when they are folded.

Joints Cracks that open up because of the deformation. These can be strike joints if they are parallel to the axis, or dip joints if they are at right angles to it. Joints may form throughout the rock parallel to the axial plane. Often these tend to fan out,

especially in beds of coarse sandstone.

Dip and Strike These were described on pp. 46–47.

Puckers or parasitic folds These form when very fine beds, such as shale, deform on a small scale as the fold forms.

The symbols are those used on conventional geological mapping.

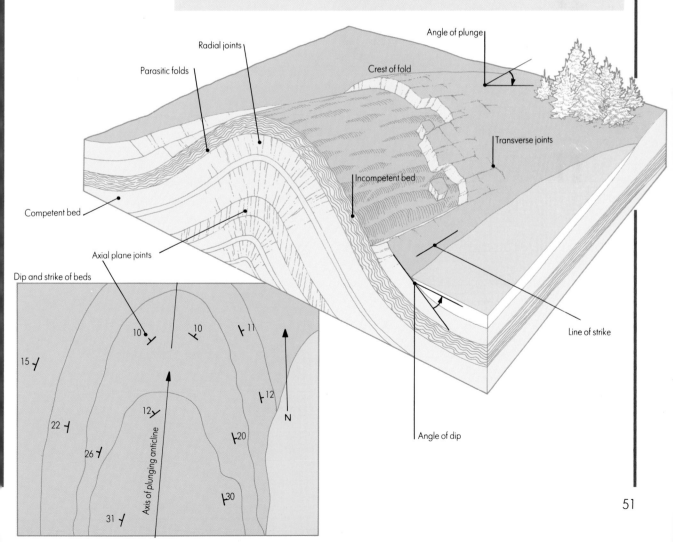

Radial joints

Parasitic folds

Crest of fold

Angle of plunge

Transverse joints

Incompetent bed

Competent bed

Axial plane joints

Dip and strike of beds

Line of strike

Angle of dip

Axis of plunging anticline

N

Faulted rocks

Sometimes rocks do not bend. Instead they break, and the rock masses move in relation to one another. This action is called faulting. Faults come in many different forms.

A dip-slip fault is produced when the movement is vertical, without any sideways component. The crack or the plane along which the fault moves does not itself need to be vertical, but is usually inclined. In a normal fault, one block has slipped down in relation to the other, down the inclined fault plane. This is produced by tension, as the rocks are pulled apart. A reversed fault or a thrust fault is one in which one block appears to have moved up the fault plane in relation to the other. This is caused by compression.

A strike-slip fault or a lateral fault is one in which the movement is predominantly horizontal. These can be further defined by the movement that has taken place – either a left lateral fault or a right lateral fault depending on which way the opposite block appears to have moved.

More often the fault is an oblique one, in which there have been both horizontal and vertical movements.

When a block moves vertically downward between two faults, the structure is called a graben. If this forms a topographical feature on the surface it is a rift valley. If a block is left upstanding as the rock masses at each side are downfaulted, the result is a horst. A geomorphological feature so produced is called a block mountain.

However, the faults do not always show themselves at the Earth's surface as hills, valleys, cliffs and so on. If the faults are very old, the whole area tends to be eroded so much that no difference in

elevation between the faulted blocks can be seen. This may pose difficulties when analyzing faults, particularly in dipping strata. It may not be obvious if a particular fault is a dip-slip fault, a strike-slip fault or an oblique-slip fault. We may have to look for other features, such as the presence of veins or intrusive igneous rocks, to estimate the displacement.

Faults can be very small, with a throw of only a few inches or centimeters. However, because of the huge-scale movements of plate tectonics, some can be thousands of miles or kilometers long. The San Andreas Fault of western North America runs for at least 807 miles (1300 km) and is caused by the relative movements of the Pacific Plate and North American Plate. It is in reality a swarm of right lateral faults, mostly parallel to one another. Movement along it has been in the region of 10 miles (16 km) over the last two million years. It is the movements along faults such as these that cause earthquakes.

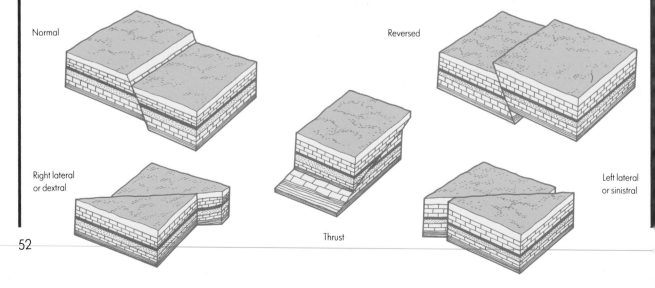

Normal

Reversed

Right lateral
or dextral

Thrust

Left lateral
or sinistral

Left The complex fracturing of this outcrop in Iran shows all kinds of faults, including normal faults **left**, a reversed fault **bottom center**, and a graben **top center**.

Right and below right
As in a fold, we can examine various features associated with a fault in the field, and most of these will give us some idea of the Earth forces that caused the disruption to the rocks – for example, whether the forces were tensional or compressional, and the direction in which they operated.

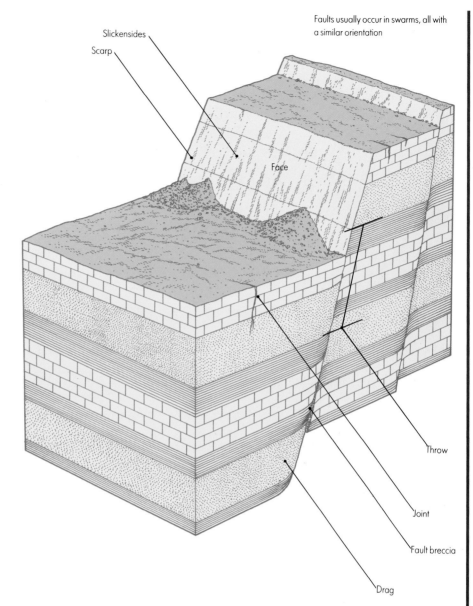

Faults usually occur in swarms, all with a similar orientation

Slickensides

Scarp

Face

Throw

Joint

Fault breccia

Drag

Things to look for in a fault

The fault shown is a normal dip-slip fault.
Throw or offset The distance moved by the fault, only measurable where different beds can be matched up.
Face The surface along which the fault has moved.
Slickensides Polished

scratch marks showing where one block has moved across the other.
Scarp The topographic feature produced at the surface if erosion has not worn it all flat.
Fault breccia A mass of rocky material broken and crushed by the movement

of the fault. In extreme cases this may form the metamorphic rock mylonite.
Drag Often the beds at each side of the fault are distorted and folded in the direction of the fault movement. This is known as drag.

Joint A break in the rock along which there has been no relative movement is called a joint.

SMALL-SCALE STRUCTURES

Folds and faults are the most obvious features produced by the deformation of rock structures by tectonic movement. There are, however, all sorts of others, just as interesting and worth looking out for. Here is a selection.

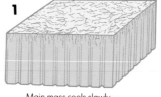

Top cools quickly – jointing irregular

Cracks form across each line of tension

1

2

Main mass cools slowly – well-defined hexagonal columns

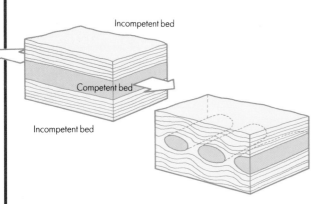

Incompetent bed

Competent bed

Incompetent bed

Columnar jointing

(*above 1–3*) As a lava flow cools, it contracts to take up a smaller volume. The contraction takes place toward a number of random centers in the mass. These centers are evenly distributed and, as a result, the whole structure splits into a number of vertical columns. These columns tend to be hexagonal, for the same geometric reasons

Boudins (*above*) When a competent bed (see p. 51) surrounded by incompetent beds is subjected to a stretching strain, it may break up into sections. The incompetent material may then squeeze around it and fill in the spaces, and in cross section (*right*) the result may look like a string of sausages.

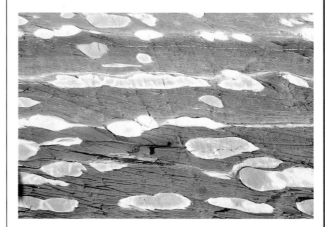

Augens (*below*) Similar deformations can affect the new minerals produced in metamorphic rocks. When a rock is subjected to such pressures and stresses that it metamorphoses into gneiss (see pp. 116–117), new crystals, such as garnet, may form. As the rock continues to strain, the crystals may rotate and open up cracks at each side. These may fill with other minerals, such as quartz, and produce an eye-like structure called an augen.

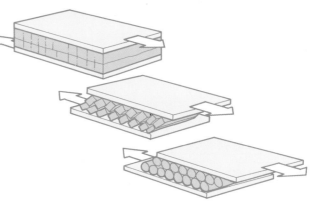

Mullions (*above*) Competent beds subjected to great stress may split up into prisms orientated at right angles to the stress. The stress may then rotate them against one another, grinding off their corners and turning them into cylindrical shapes.

3

Veins (*right*) Joints that are caused by tension, rather than compression, may open up. The spaces so formed tend to be filled with minerals deposited by groundwater, usually quartz but sometimes calcite. The veins so formed can be very visible when the light-colored mineral contrasts with a dark-colored rock.

that honeycomb is hexagonal – it allows the largest number of units in a given space. A second type of jointing may then split each column into regular sections, so that they resemble stacks of hexagonal prisms. Many landmarks and beauty spots, such as the Giant's Causeway in Northern Ireland and Devil's Tower in Wyoming, are a result of such columnar basalts.

Dendrites (*below*) Often a joint is so narrow that water can only seep along it with difficulty. It does not then deposit its minerals evenly along the face, but as a kind of a gradual growth. Manganese oxides are often

deposited in this way, producing a branching structure that looks like a fossil plant. This sometimes happens inside a silica mineral producing the gemstone called moss agate.

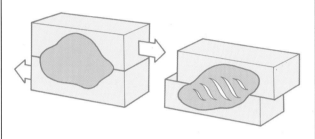

***En échelon* fractures** (*above*) When two masses of rock shear past one another, the stresses set up between the two open up joints. These joints lie along the plane of the shear direction, but are at an angle to it. You can observe the same effect by spreading clay, or pastry, over two blocks of wood placed side-by-side and then moving the blocks past one another. *En échelon* joints appear in the clay or pastry.

If the stresses continue, the rock between the joints collapses and forms a fault breccia. Otherwise the angled joints fill with deposited minerals.

These are all features that affect rocks after they have been formed. Sometimes they make such a mess of the original rock that it is very difficult to work out the history of the area. However, a useful rule is that, if structure A cuts across structure B, then structure A is younger than B. For example, if a vein cuts across a set of mullions and then is stopped suddenly by a fault, we can deduce that the mullions were formed first, then the vein injected, and then the whole system faulted.

IGNEOUS STRUCTURES

Molten rock extruded through cracks and filling crevices solidifies into an intrusive igneous rock (pp. 32–33). The results form characteristic rock structures.

On the large scale, huge volumes of magma can rise into the continental rocks, usually melting and assimilating the continental material, and solidify in vast masses deep underground. This usually happens in the core of folded mountains, and the enormous structure so formed is called a batholith. Granite is the rock most usually found in a batholith. When the mountain chain is eroded away to its deep interior, the batholith will appear at the surface, forming broad tracts of moorland with prominent granite outcroppings. The moors of southwest England are the surface expressions of a vast batholith that lies beneath Devon and Cornwall and reaches out to the Scilly Isles.

In general, igneous rocks are harder than the rocks into which they are injected. When the landscape is eroded away, they tend to stick out as distinct landscape features – sills as cliffs jutting out of a hillside, and dikes as wall-like structures cutting across the scenery.

Right Magma seeks out planes and lines of weakness in the rocks, and thrusts its way through them. When these lines of weakness are the bedding planes of sedimentary rocks, and the igneous rock is emplaced as a sheet between the beds, the structure so formed is called a sill. A laccolith forms when a mass of molten material gathers at a point and domes up the strata above, like a septic pimple.

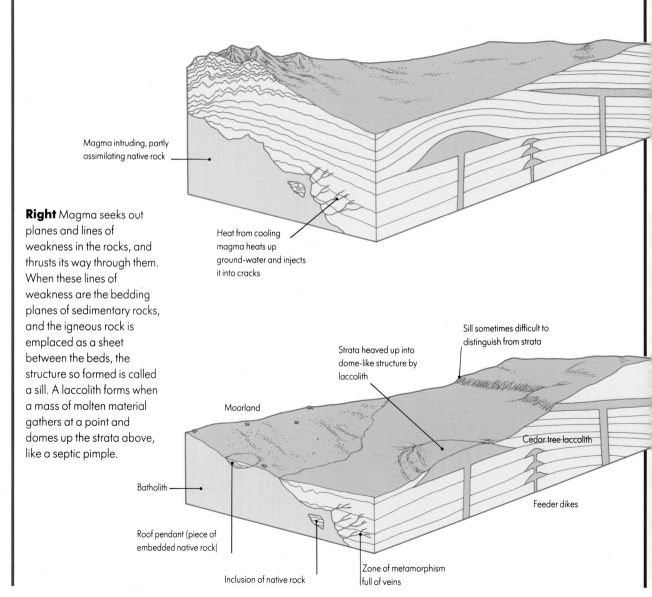

Magma intruding, partly assimilating native rock

Heat from cooling magma heats up ground-water and injects it into cracks

Strata heaved up into dome-like structure by laccolith

Sill sometimes difficult to distinguish from strata

Moorland

Cedar tree laccolith

Batholith

Feeder dikes

Roof pendant (piece of embedded native rock)

Inclusion of native rock

Zone of metamorphism full of veins

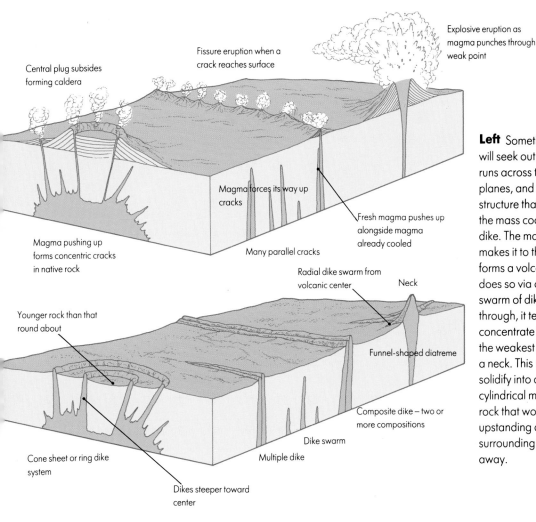

Central plug subsides forming caldera

Fissure eruption when a crack reaches surface

Explosive eruption as magma punches through weak point

Magma forces its way up cracks

Fresh magma pushes up alongside magma already cooled

Magma pushing up forms concentric cracks in native rock

Many parallel cracks

Radial dike swarm from volcanic center

Neck

Younger rock than that round about

Funnel-shaped diatreme

Composite dike – two or more compositions

Cone sheet or ring dike system

Dike swarm

Multiple dike

Dikes steeper toward center

Left Sometimes the fluid will seek out a crack that runs across the bedding planes, and the solid structure that is formed as the mass cools is termed a dike. The magma that finally makes it to the surface and forms a volcano usually does so via a dike, or a swarm of dikes. As it bursts through, it tends to concentrate its force along the weakest point and form a neck. This may eventually solidify into a vertical cylindrical mass of igneous rock that would be left upstanding as the surrounding volcano is worn away.

SILL OR BURIED LAVA FLOW?

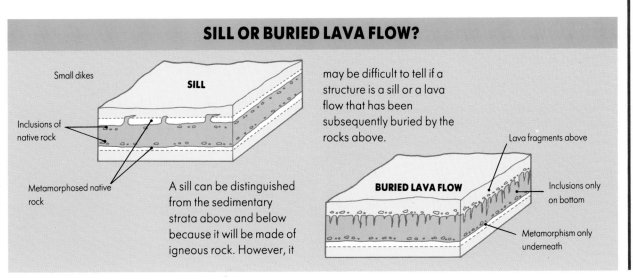

Small dikes

SILL

Inclusions of native rock

Metamorphosed native rock

may be difficult to tell if a structure is a sill or a lava flow that has been subsequently buried by the rocks above.

A sill can be distinguished from the sedimentary strata above and below because it will be made of igneous rock. However, it

Lava fragments above

BURIED LAVA FLOW

Inclusions only on bottom

Metamorphism only underneath

57

THE EARTH'S STORYBOOK

An ocean bed can become a mountain, then a plain, with a succession of different plants and animals. These events can all be read in the rocks.

HISTORY IN THE ROCKS

Sedimentary rocks are mute witnesses to their own creation. Each one could tell the story of how it formed and the conditions of the Earth's surface at the time. The nature of the rocks themselves, as we have seen (pp. 32–37), gives us part of the story. Limestones form in carbonate-rich seas; rock salt forms under conditions where seawater is exposed on shallow sub-marine surfaces and evaporated; mudstones and shales form in muddy water; sandstones form in deserts or on sandy river beds; etc.

However, it is the structures within the rocks that give us the details about the processes of their formation. Here are a few structures to look for.

Tool marks A sea current may carry a shell or a fragment of rock along, bouncing it off the bottom and dragging it across the sea bed. The marks left can give us an indication of the direction of the current.

Usually it is not the mark itself that is preserved, but the cast made as the subsequent sediment fills the hole. When the rock is finally exposed, the fine material in which the marks were made tends to be very fragile and is eroded away quickly. The material above is coarser and sturdier – usually the bottom of a graded bed – and preserves the impression longer. This is why we more often see dinosaur footprints as three-dimensional raised surfaces on the rock outcropping than we do the original depressions made by the animal.

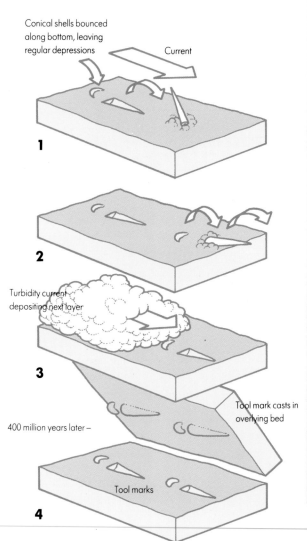

Conical shells bounced along bottom, leaving regular depressions

Current

1

2

Turbidity current depositing next layer

3

400 million years later –

Tool mark casts in overlying bed

Tool marks

4

Salt pseudomorphs

When salt-water puddles dry out, they leave the salt as crystals. Sometimes the crystals may be fairly large, forming distinct cubes. When water returns to the area, the salt dissolves and is washed away, but may leave behind cubic depressions which can then be filled up with later sediment. When these crystal shapes are preserved as casts in stone, they are called pseudomorphs – false shapes (see box).

Ripple marks We have all seen wave marks left in the sand at low tide. The ripple marks, an inch or two high, that are so formed can be preserved in the bedding of the resulting sandstone. The presence of such ripple marks in a sandstone indicate that the sediment was deposited in shallow water, with waves breaking.

Recording sedimentary structures

Collecting sedimentary structures may be neither practical nor, in these days of conservation, desirable. Yet there is a technique that can be used on certain small-scale structures that stand out in relief in the rock exposure. That is the technique of rubbing often used to reproduce the sculpting on graves and other raised surfaces.

Take a piece of soft paper big enough to cover the structure. Wallpaper lining paper is ideal. Tape it down to the rock at the edges. Now take a block of heel-ball – a waxy compound originally used by shoemakers, but nowadays used more often for making rubbings – and rub it lightly over the surface, with an even pressure, as if shading the whole paper with a soft pencil. The image will then be preserved for later study.

Hollow squares represent hopper crystals – crystals with stepped faces.

Three-pronged shapes are the corners of cubes sticking up.

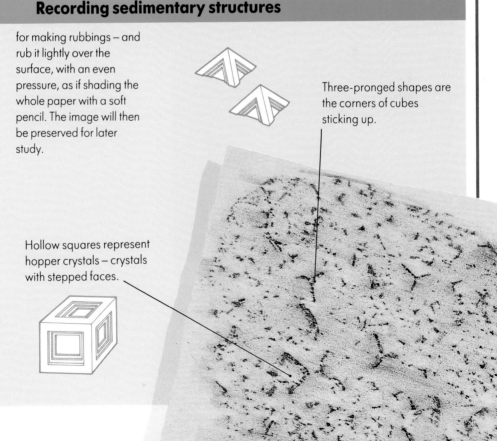

1 Most sediment deposited over end of tongue

River current

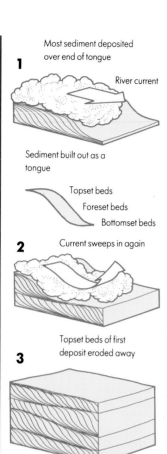

Sediment built out as a tongue

Topset beds
Foreset beds
Bottomset beds

2 Current sweeps in again

Topset beds of first deposit eroded away

3

Final result

Current bedding

River sandstones can be identified by the presence of current bedding. River sand is deposited as a tongue. Successive deposits of sand build out from the end of the tongue, in S-shaped beds – a thin layer along the top (the topset), a thick layer at an angle at the front (the foreset), and a thin layer spreading forward on top of the one before (the bottomset). Then, when the current changes, the topset and the top part of the foreset are swept off, and the tongue of sediment builds out again on top of what is left. The resulting sequence of beds shows curved structures that represent the bottom parts of the S-shapes, concave in the direction that the current is flowing.

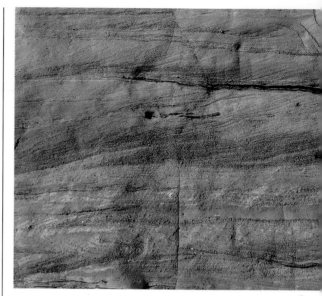

Curre

Slump bedding

Sometimes sandstone bedding seems to break down completely, producing a contorted mass. This is the result of sandstone beds remaining quite fluid, and distorting with the pressure of more sand accumulating on top, or of continuing current action. Slump bedding can be thought of as fossil quicksand.

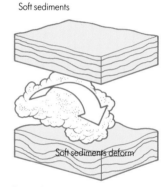

Soft sediments

Soft sediments deform

Current deposits a mass of new sediment on top

Dune bedding

Parts of original sediment forced up as sedimentary dikes and dragged over as flame structures

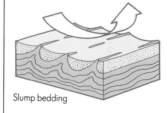

Slump bedding

Flame structures

When a mass of sediment is deposited suddenly on a soft layer, the mass may sink into it. The softer material may then squirt upward and be dragged away by the current. The result, in cross section, has a very sinuous, tapered appearance reminiscent of flames.

Dune bedding

Exactly the same thing happens when desert sands move over one another as shifting dunes. However, whereas tongues of river sediments may be a dozen or so inches thick, sand dunes can be tens of yards. The thing that distinguishes dune bedding from current bedding in a sandstone is the sheer scale of the structure. structure.

Mudcracks Mud exposed to the sun dries out and shrinks. As it does so, it cracks into polygonal-shaped slabs. The cracks may be filled in by a subsequent deposit of sand. Polygonal cracks such as these, or the ridges formed in the overlying sandstone where it has been molded in the cracks, show that there was an environment in which mud could dry out completely.

Rain pits In dry areas the occasional raindrops may splash pits into exposed fine sediments. When turned to stone, these can be regarded as fossil weather.

Coarseness A fast current can carry heavier fragments than a slow current. As a rule, the coarser sedimentary rocks, such as conglomerates, were deposited by faster-moving water than were the finer shales. Sometimes a current slows and stops. When this happens the heaviest material that is carried along is deposited first, and then the lighter on top of it. The result is a bed of, say, sandstone with coarse grains on the bottom, grading to finer grains at the top.

We often get this in deep sea sediments where currents sweep material from the continental shelf down to the depths, fading out as they spread over the flat ocean plain. Such currents are called turbidity currents and the bed produced by them is called a turbidite.

A sandstone made up of different-sized fragments shows where the current came to a halt suddenly and may show turbulent conditions.

Roundness As rock fragments are swept along by a current, they tend to have their edges knocked off, and they become increasingly spherical.

When we see jagged fragments in a sedimentary rock we can tell that the fragments have not been transported very far. Well-rounded fragments, on the other hand, have been polished a great deal before coming to rest.

The roundest sand grains are found in desert sandstones. They have been carried about in winds, duststorms, and shifting sand dunes for millennia before finally being buried and turned to rock.

Sole marks When a turbidity current sweeps over an area of marine deposition, the soft fine sediment deposited at the top of the previous bed is disturbed and scoured into hollows. These hollows are reflected in the structure of the coarse bottom sediment of the subsequent bed and preserved in the final rock – on the sole of a graded bed. They may be in chains of crescent-shapes with the hollows pointing upstream, in which case they are called flute casts.

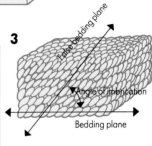

Flattened stone carried along by strong river current tends to rise at front

Stones settle, sloping upstream

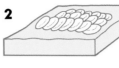

False bedding plane
Angle of imbrication
Bedding plane

Imbrication When a flat pebble is swept along by a current, the hydrodynamics involved produce an eddy below the leading edge and above the trailing edge. As a result, the pebble tends to settle on the bottom at an angle, sloping upstream. The result is an imbricated structure. When this happens to a large mass of pebbles in a conglomerate, the angle of imbrication may be confused with the bedding plane. Be aware.

Flame structures

Flute casts

SEDIMENTARY SEQUENCES

Stand on the wharf of a river port and watch the river flow down to the sea. It may be that a few hundred years ago this port was on the sea itself. The sediments being washed down by the river may have been accumulating at the river mouth for hundreds of years, building up the land. If the landscape can change like that in a few centuries, imagine the changes during the vast sweep of geological time!

Such changes are recorded in the rocks. A typical rock sequence from early Carboniferous times may run, from the bottom up (it is geological convention to deal with rock strata from the bottom up, since this reflects the sequence of deposition), as follows. There may be a bed of limestone containing the fossils of marine creatures. Above this, there may be a bed of shale, still with marine fossils. Then may come a bed of sandstone – several beds, in fact, that may build up into a thick sequence. This sandstone may have current bedding in it. Toward the top the sandstone may become very pale, and have fragments of carbon in it. Above this, there may be a bed of coal. Then the limestone may appear again, followed by the shale, and so on. This sequence may be repeated many times.

The fact that we described this sequence from the bottom upward reflects a fundamental law in geological study. The Law of Superposition states that in any undisturbed sequence of rocks, the oldest rocks are at the bottom. It seems quite obvious now, but when Hutton formulated it in the late 18th century it ran counter to much of the since discredited Neptunian theories.

The term "undisturbed sequence" is important here. A sequence of sedimentary rocks may be completely inverted, as in the arm of a recumbent fold (pp. 50–51). Then we have to look at the structures within the beds to see which way up it should be – graded beds grading upward, current bedding curving the right way, and so on. This is a procedure that field geologists term younging.

Another important rule is the Principle of Lateral Continuity: When we find the same rocks outcropping in different places, we can assume that they were once continuous and that they have been eroded away from the intervening region.

Analysis of a cyclical sequence

Where rocks are formed from beds deposited in a delta, the beds often follow a particular pattern, which can tell the story of their development. The sequence may start with limestone, showing the presence of clear shallow sea water. Above this there will be a deposit of shale, mudstone, or clay, formed from mud particles drifting in from a nearby river. Then will come a bed of sand, probably current bedded, river deposits on the site. The top of the sandstone may be pale and full of root fragments and will be followed by a bed of coal. This indicates that the bed of sand formed a sandbank above the level of the water and supported vegetation. The plants leached the nutrients from the sand (the pale colour)

Left and below This stream bank on the Scottish-English border shows a typical early Carboniferous sedimentary sequence. At the base (**1**), mostly beneath the rubble, is a thick bed of mudstone, deposited in muddy waters. Above this is a protruding bed of hard clay (**2**) formed from finer particles deposited in shallow cloudy water. Then comes a bed of softer clay (**3**), which formed under similar conditions but is dark with plant fragments and is less compact. This must have been formed in very shallow water for above it is a bed of coal (**4**), the outcrop of which is stained red with iron from the other rocks. The coal is overlaid by more soft clay (**5**), showing that the vegetation was flooded by shallow still water. Above this comes a sequence of thinly bedded sandstones (**6**) jutting out of the cliff face in individual beds. These were deposited in a river with a variable current. Quieter waters followed giving another bed of soft clay (**7**), and then more sand settled to produce a thick bed of massive sandstone (**8**) forming the overhang at the top of the cliff.

and left their roots behind, producing a rock called seat-earth. After the coal may come another bed of shale or sandstone, or even of limestone as the vegetation was flooded by the river or the sea, and the whole sequence starts again as the sandbanks build up. Many of the delta deposits of Carboniferous times in Europe and North America show this cyclical sequence as the sea came and went, repeatedly flooding the deltas as the region slowly subsided. Rock sequences may consist of thousands of cycles based on this general pattern, and may produce the thick beds of coal that were the mainstay of the Industrial Revolution.

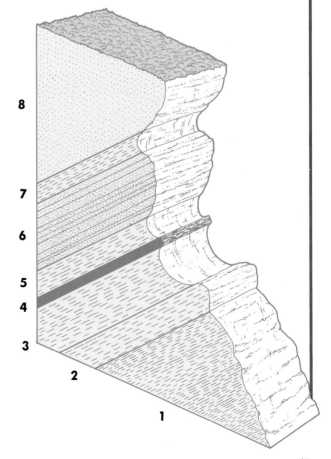

UNCONFORMITIES

When a sequence of rocks is lifted above sea level, the wind and the rain act almost immediately to wear it down again. Erosional forces break up the exposed rocks and transport the fragments away to form new sedimentary rocks. The landscape is eventually worn flat, and the sea may cover it again.

When this happens a new sequence of sediments is built up on the eroded surface and these eventually become sedimentary rocks. The break between these new rocks and the old rocks below is called an unconformity.

In most occurrences there is a bed of conglomerate (see pp. 34–35) immediately above the unconformity. This represents the remains of a shingle beach that formed on the eroded surface as the sea advanced across it. The advance of the sea is called a transgression. When the sea retreats it is called a regression. However, a regression is not easy to recognize in a rock sequence. It leaves sediments exposed to the weather, and these are then eroded, destroying any evidence of the sea's retreat.

The presence of an unconformity is important in dating the sequence of the rocks. We have seen how we can date the geological events of an area by observing which igneous sstructure cuts what, and which fault breaks through what sequeence. If an unconformity cuts across a dike, we can deduce that the dike was emplaced first and the rock sequence was later eroded to produce the unconformity.

The study of the sequence of rock types, and of the stories they tell, is called stratigraphy. When applying stratigraphic techniques over a large area we can sometimes find the distribution of land and sea and the various environments at a particular period of geological time. This study is called paleogeography.

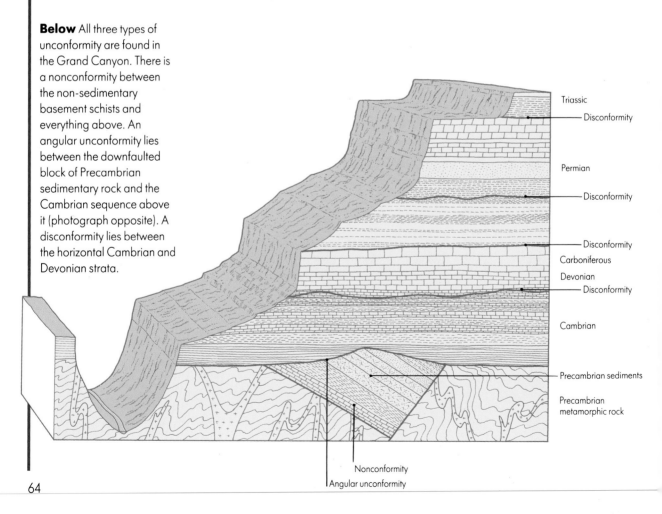

Below All three types of unconformity are found in the Grand Canyon. There is a nonconformity between the non-sedimentary basement schists and everything above. An angular unconformity lies between the downfaulted block of Precambrian sedimentary rock and the Cambrian sequence above it (photograph opposite). A disconformity lies between the horizontal Cambrian and Devonian strata.

Triassic
Disconformity
Permian
Disconformity
Disconformity
Carboniferous
Devonian
Disconformity
Cambrian
Precambrian sediments
Precambrian metamorphic rock
Nonconformity
Angular unconformity

1

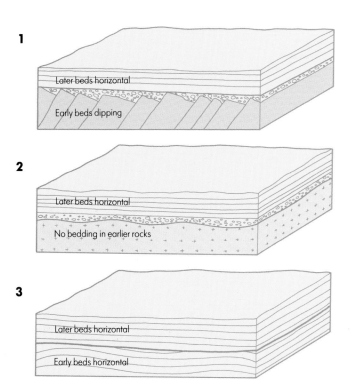

Later beds horizontal

Early beds dipping

2

Later beds horizontal

No bedding in earlier rocks

3

Later beds horizontal

Early beds horizontal

Above There are three sorts of unconformity:

1 Angular unconformity This is the easiest to spot. The beds beneath lie at an angle to those on top.

2 Nonconformity Another easy one. There are no beds below. The later deposits were laid down on an igneous mass with little internal structure.

3 Disconformity In this there is no clear division between the lower and the upper beds. The lower beds were eroded down but without being disturbed first. Sometimes the only way we can tell the presence of a disconformity is by recognizing fossils from quite different ages above and below it.

Symbols for rock types

The symbols used to represent rock types on a geological diagram, or a map.

 Limestone

 Dolomite

 Dolomitic limestone

 Shaly limestone

 Sandy limestone

 Shelly limestone

 Conglomerate

 Breccia

 Sandstone

 Current-bedded sandstone

 Siltstone

 Shale

 Mudstone and clay

 Anhydrite

 Coal with seat earth

 Coarse igneous rock

 Fine-grained igneous rock

 Lava flow

 Slate

 Gneiss and schist

65

FOSSILS

Animals and plants were not always as they are today. The ancestors of the modern flora and fauna were sometimes quite different from their offspring. We can tell the history from the fossil remains that they have left behind them in the rocks.

A fossil can be preserved in any one of a number of ways.

Sometimes, but not often, the entire organism may by preserved unaltered. We see this in the bodies of mammoths preserved in the frozen mud of the tundra. We also see it in the bodies of insects trapped in amber, the fossil resin of trees.

More usually only the hard part of the organism remains. Sharks' teeth in the Tertiary clays of eastern England and the bones of saber-toothed cats in the tar pits of Los Angeles consist of unaltered original material.

Usually the original material is broken down chemically so that there is little left. The black shapes of fern leaves we find in Carboniferous shales consist of the leaves' original carbon. This process taken to an extreme gives us coal.

The original material may have been replaced, molecule by molecule, by mineral substances, producing a fossil that retains the original cellular structure but in a different material – a process called petrification. The wood of Petrified Forest in Arizona still has its microscopic structure but is now made of silica. A mold is formed when the organism is dissolved away entirely, leaving a hole in the rock the exact shape of the organism.

When a mold fills with minerals deposited from ground water it produces a cast – a lump of mineral that has the shape of the original organism but shows no internal structure.

Similar to a cast is a bioimmuration. In life one sedentary organism may grow over another, completely engulfing it. The imprints of soft-bodied organisms like bryozoans can be found on the undersides of oyster shells where they have grown over the underlying rock.

Finally, a fossil need contain no part or shape of the original organism. Trace fossils are footprints,

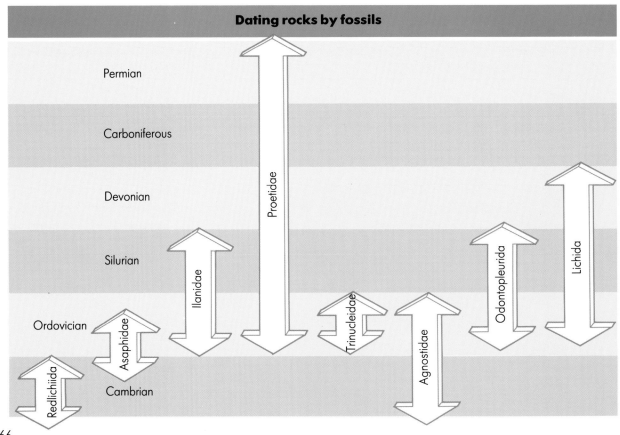

Dating rocks by fossils

A fossil fern leaf consists of a film of the original carbon **above**. Masses of plant material such as this give rise to coal. A graveyard of fossil fish **left** shows where some disaster overcame a living community, probably due to the drying out of a desert stream.

worm burrows, feeding trails, and the like – evidence of where an animal was living.

We study fossils for a number of different reasons. Perhaps the most common is to appreciate the wonderful panoply of the life of the past. We can more fully appreciate the life of the world today if we know what has preceded it.

Studying fossils has enabled us to build up a pattern of evolution, each stage of which we can apply to some particular time in the geological past. We know, for example, that the multi-legged sea animals called trilobites were abundant from

Cambrian to early Devonian times – 590 to 408 million years ago – and continued until the Permian – up to 248 million years ago. Therefore, if we find the fossil of a trilobite in a rock we can say that the rock is most likely Cambrian, Ordovician, or Silurian in age, although it may be Devonian, Carboniferous, or Permian. If we can identify the trilobite that will be better still. For example, we know, from scientific inquiry over the last 200 years, that the agnostids – a distinctive family of very small trilobites – lived in Cambrian and Ordovician times. We also know that another

The ammonite fossil **left** shows where the interior of the shell filled with sediment and formed a cast. Some of the altered shell material remains on the outside. The insect in amber **below** illustrates the unusual occurrence in which the entire organism is preserved unaltered.

distinctive family – the lichids – lived from Ordovician to Devonian times. If we find representatives from both families in the same rock, we can pinpoint the rock as being Ordovician in age, that being the only period in which the ranges of the two families overlapped.

If we can identify fossils down to the species and subspecies level, they can be even more useful in dating rocks. Ammonites – animals like squids in coiled shells – were abundant in Triassic, Jurassic, and Cretaceous seas – 248 to 65 million years ago. The Jurassic rocks of southern England, in particular, are divided into several dozen time zones based on the species of ammonite found in them.

Nowadays detailed chronologies have been developed based on microfossils including forami-niferans, nanoplankton, and diatoms – fossils of tiny plants and animals most of which can only be seen with the help of a microscope. The advantage of this is that small fragments of rock brought up from boreholes by oilwell engineers can be studied by paleontologists, and the rocks penetrated by the drill can be dated.

The fossils used for dating rocks are called index fossils. Two criteria must be satisfied to make a fossil useful for chronologic applications. The organism must be geographically widespread – such as a creature that inhabits all the seas of the world. This ensures that fossils can be found the world over. Second, it must have a high rate of evolution, so that after a few million years it has produced a recognizably new species.

Fossils have other uses. Facies fossils are indicators of the environment in which they lived. For example, the shellfish *Scrobicularia* lives buried in the oxygen-poor mud of estuaries – the kind of conditions that clog up the gills of most shellfish. So if we find a shale with fossils of *Scrobicularia* we can tell that the shale formed from the oxygen-poor mud of an estuary. This is useful because it can guide the oil geologist to the type of rocks which are likely to have formed under the conditions suitable for oil production.

Fossils are rarely found in isolation. Usually they are found as an assemblage. Two kinds are recognized by the geologist. The first is the life assemblage or biocoenosis. In this the creatures are preserved just as they lived. Delicate animals like sea lilies are still articulated, bivalves have their shells together, worm burrows and feeding trails are undisturbed. It may come about when the whole area is engulfed in mud, killing everything and burying it instantly. Such an assemblage is valuable in that it shows how the animals lived in relation to one another. The other assemblage is the death assemblage or thanatacoenosis. In this the skeletons are disarticulated and scattered, bivalves are broken apart and may be aligned in the direction of current. Such an assemblage can indicate the currents that flowed at the time, but is not too useful otherwise. The currents may have brought in organisms from other environments, or even washed fossils out of preexisting rocks.

Make your own fossils

To make a mold fossil, you can pour a layer of plaster of Paris into a shallow tray. Then, before it has dried, push a clay model of a shell into it. Once it is dry, lightly grease the surface and pour on another layer, completely covering the shell model. Once the whole block is dry, you can remove it and split it apart along the joint. If you now remove the shell model, the hole left represents the mold. To make a cast, you can take one half of your plaster block and cut a channel from the outside to the hollow. Then put the two halves back together and pour liquid latex down the hole made by the channel. Once the latex is dry you can separate the pieces once more and find that the latex has taken the precise shape of the original clay model.

Trace fossils can be done the same way. Pour a layer of plaster of Paris and press a fern leaf into it. When that has set, you can grease the surface of the plaster and pour in another layer. Once it is dry you can split it apart. You will see your original print in one half, and in the other half there will be an upraised image of the leaf.

An insect trapped in amber can be simulated by using a liquid resin hobby kit. Mix some liquid resin and hardener, following the instructions with the kit, and pour it into a mold. When it has begun to harden, place a dead insect on it and pour on another layer. Once the resin has hardened, your insect will last indefinitely.

EROSION AND GEO-MORPHOLOGY

The relentless movement of the Earth's surface plates causes landscapes to be uplifted and mountains to be heaved skyward; no sooner is an area of rock raised above sea level than the wind and the rain, the ice and the rivers, the sun and gravity all act to break it down once more. The shape of landscapes (geomorphology) can be seen as the temporary balance between these forces.

Natural decay

Visit a graveyard, preferably an old one. The recent gravestones will be fairly clean, and their inscriptions will be fresh. However, the older ones will be worn and decayed, and the oldest will be almost indecipherable. See which is the oldest date that you can read.

Most of the older monuments will be of the same stone – probably a local stone – and you will see how quickly that stone erodes once it is exposed by looking at the inscribed dates. This tends not to work for more modern gravestones, since they may be sculpted from a wider variety of materials.

If the graveyard consists of monuments made from many different kinds of stone, you will find that some kinds weather more quickly than others. The metamorphic rock marble, for example, decays surprisingly rapidly compared to the sedimentary limestone from which it is derived.

Different climates impose different rates of weathering. Hot damp conditions tend to rot any material quicker than cold dry ones. However, under any conditions, two types of weathering can be distinguished – physical weathering and chemical weathering.

Physical weathering

In the first of these it is the mechanical effects of wind and torrential rain, frost, and animal movements that produce the erosion.

Perhaps the most important factor in higher latitude and altitude rock erosion is frost. Rainwater seeps into pores and cracks in the rock. When this freezes to ice it expands by about nine percent of its volume, forcing open the pores and cracks – in the same way that water pipes become damaged if allowed to freeze in the winter. More water can then seep in and when this freezes even more force is applied. The pressure applied can be about a hundred times greater than the pressure of air in a car tire. Eventually the exposed rock disintegrates into rubble, shattered angular blocks of which lie in long slopes termed scree sweeping down from jagged splintery rock faces.

In hot dry climates the difference in temperature between day and night can have a destructive effect. Rocks expand in the heat and contract in the cold. When this happens to exposed surface layers, they tend to become separated from underlying layers. This is most obvious when the rocks are distinctly bedded, with planes of weakness parallel to the surface. When it happens on massive rocks such as granite, the result is that the rock erodes in curved slabs, a process that is given the technical term exfoliation, or the more descriptive term onion-skin weathering. This action is undoubtedly aided by what is known as pressure release, as the rocks expand after an overburden is eroded away. Chemical weathering may also have a part to play in this complex business.

The real mover in arid landscapes is the wind. The wind can pick up dry particles and hurl them along, blasting them against exposed rocks and

Scree slopes, such as those in the English Lake District **above**, are produced by physical weathering. Water in pores and cracks in the rock expands as it freezes, forcing the rocks apart. Biological erosion takes place as tree roots seek out joints in the rock and then split the rock as they grow, as in this Philippine example **far left**. Onion-skin weathering, where a rock spalls away layer by layer, as here in Tanzania **left**, is caused by both physical and chemical erosion in arid climates.

71

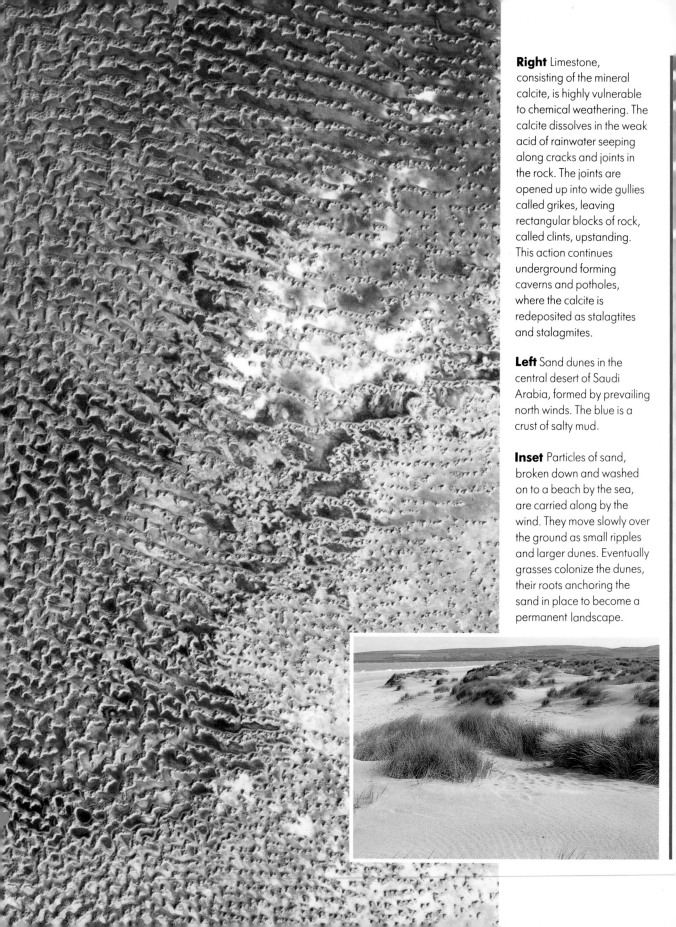

Right Limestone, consisting of the mineral calcite, is highly vulnerable to chemical weathering. The calcite dissolves in the weak acid of rainwater seeping along cracks and joints in the rock. The joints are opened up into wide gullies called grikes, leaving rectangular blocks of rock, called clints, upstanding. This action continues underground forming caverns and potholes, where the calcite is redeposited as stalagtites and stalagmites.

Left Sand dunes in the central desert of Saudi Arabia, formed by prevailing north winds. The blue is a crust of salty mud.

Inset Particles of sand, broken down and washed on to a beach by the sea, are carried along by the wind. They move slowly over the ground as small ripples and larger dunes. Eventually grasses colonize the dunes, their roots anchoring the sand in place to become a permanent landscape.

gradually wearing them down. Most of this action takes place close to the ground where the sand particles are being bounced along. A common result is a rock that looks like a mushroom, with a broad head and a narrow stalk that has been polished away by the sandblasting. Another result is the dreikanter, a stone that has been polished on three sides. A stone lies on the ground. Sand driven by the prevailing wind wears down one side. It becomes unbalanced and so topples over, exposing another side to be polished.

An effect of all this blasting and polishing is that more sand is produced, which adds more force to the sandblasting. Desert sand is moved over the desert floor as slow-motion waves, ie dunes.

A final aspect of physical weathering is the role of living things. Trees growing in soil sink their roots into the bedrock. These roots follow cracks and expand them, splitting the rock open. Certain shellfish can even burrow into rocks, breaking them down.

Nowadays the most potent biological agents of physical erosion are people. Inappropriate farming methods can destroy topsoil. Well-worn footpaths and tracks become incised into the landscape, especially where the native rock is something soft like loess. There is hardly a corner of the globe where human beings have not been a major influence in landscape formation.

Chemical weathering

The gentle rain can often be quite harsh. It may be very acidic, either because it has dissolved acidic industrial gases, or more probably because it has dissolved carbon dioxide from the atmosphere and formed carbonic acid.

Certain minerals are susceptible to attack by this airborne acid, notably feldspar and calcite. Granite consists mostly of quartz, feldspar, and mica. In wet climates the feldspar can react with atmospheric acid and decay into clay minerals. This loosens the other minerals in the rock and they fall out. As a result granite landscapes have china clay quarries, and white beaches of quartz and mica sand by the sea.

Limestone, which consists largely of calcite, is also eroded by acidic rain. The water seeps along joints and cracks, dissolving their sides as it goes, and these cracks are opened up into crevices called grikes, leaving the intervening rocks as upstanding blocks called clints.

Basic igneous rock such as gabbro is also weathered in this way. The olivine is particularly susceptible, but more to water than to the acid it contains. Water penetrates along joints and attacks all sides at once. The result is that the fresh rock is eroded away. The corners are most vulnerable and become worn off so that the mass becomes a collection of spheres – hence spheroidal weathering.

The chemistry of erosion

The reactions involved in chemical weathering are quite complex, involving the production of soluble substances that are then washed away.

Water (H_2O) dissolves carbon dioxide (CO_2) and becomes carbonic acid (H_2CO_3).

The reaction with feldspar to produce clay minerals –

$6H_2CO_3 + 2KAlSi_3O_8 = Al_2Si_2O_5(OH)_4$ (clay) + $4SiO(OH) + K_2CO_3$ (soluble)

The reaction with calcite –

$2H_2CO_3 + 2CaCO_3 = H_2 + 2CaHCO_3$ (soluble)

THE WATER CYCLE AND THE EFFECT OF RIVERS

Water is in constant motion across the surface of the Earth – not just in the surging tides and crashing waves of the sea, and the flowing of the majestic rivers, but throughout the atmosphere also. The sun evaporates water from the ocean surface and the wind transports the vapor. When conditions change, such as when the temperature drops, the vapor condenses, first to droplets that form clouds, and then to drops that form rain. When the rain falls on the land, it may sink into the soil, or flow over the surface. Eventually it finds its way into streams and rivers and flows back to the oceans. This whole sequence is called the water cycle, and it has a profound influence on the life and the landscape of our planet.

The ages of rivers

Geographers and geologists acknowledge that rivers pass through three stages – youth, maturity, and old age. The classification is a little anthropocentric but it describes well the river's conditions and actions at various stages of its existence.

Naturally, all rivers differ. Sometimes a stage is left out, with a river dropping straight out of its youthful stage in the mountains to old age on the plains. Each river varies through time also, changing its stages as mountains are worn away and plains built up.

Rivers for the geologist

All these stages are of interest to the geographer. However, it is the youthful stage that is most important to the practical geologist. Here the river is constantly eroding the rocks, giving good exposures and cross-sections of the geology of the area. When the bedrock is worn smooth by water currents, any fossils contained often stand proud if they are made of a harder mineral. Waterfalls and rapids form over beds of harder rock, giving an instant indication of the layout of the geology.

In the mature stage the deposited debris will be more evident than the bedrock. This will be exposed on the outside of curves, where the current is fastest and eats into the bank. The bluffs of the valley may be cut down to the native rock but they are likely to be overgrown. Sometimes interesting rocks are seen embedded in the roots of dead trees washed downstream and deposited during floods. These can give an indication of what the geology is like in the mountains.

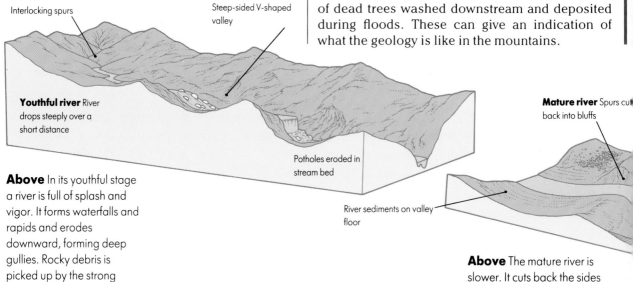

Interlocking spurs

Steep-sided V-shaped valley

Youthful river River drops steeply over a short distance

Potholes eroded in stream bed

Mature river Spurs cut back into bluffs

River sediments on valley floor

Above In its youthful stage a river is full of splash and vigor. It forms waterfalls and rapids and erodes downward, forming deep gullies. Rocky debris is picked up by the strong current and carried along.

Above The mature river is slower. It cuts back the sides of its valley, but at the same time deposits material on the valley floor, forming a plain. The river changes course continually and winds about on the plain.

Right Water is continually circulating at the surface of the Earth. It evaporates from the oceans into the atmosphere where it may drift over land. Later it may condense and fall as rain. This will run off the continents as rivers and streams, or soak through the soil and rocks. Eventually it will reach the ocean again. Along the way some will evaporate from the soil or from rivers and lakes back into the atmosphere. Some water will be drawn up into plants, where it will evaporate from the leaf surfaces. The entire process is called the water cycle.

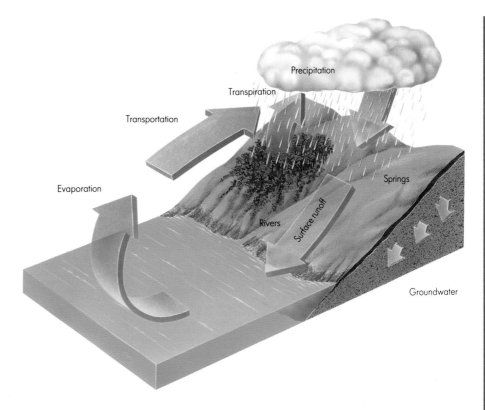

Precipitation

Transpiration

Transportation

Evaporation

Springs

Rivers

Surface runoff

Groundwater

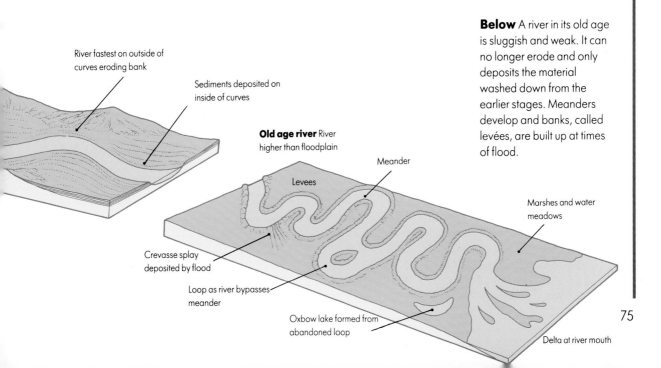

River fastest on outside of curves eroding bank

Sediments deposited on inside of curves

Old age river River higher than floodplain

Levees

Meander

Crevasse splay deposited by flood

Loop as river bypasses meander

Oxbow lake formed from abandoned loop

Marshes and water meadows

Delta at river mouth

Below A river in its old age is sluggish and weak. It can no longer erode and only deposits the material washed down from the earlier stages. Meanders develop and banks, called levées, are built up at times of flood.

75

RIVER PATTERNS

Throw water onto a rough surface. It runs off through the random hollows, skirting around the higher points, but generally flowing downhill. Rivers, in their progression from youth through maturity to old age, do exactly the same, and the hollows and high points that they follow or avoid are largely due to the geology of the area. An aerial photograph of the river pattern that forms on structureless rocks, such as a vast granite batholith (pp. 54–55), shows streams and tributaries converging in a random, but even, pattern. Such a pattern is called a dendritic drainage pattern – a reference to its similarity to the branching of trees.

The effect of geology on rivers

Dipping strata produce quite different drainage patterns. Where soft rocks alternate with hard rocks, for example shale interbedded with sandstone, the soft rocks always erode first. A river naturally tends to follow the outcrop of the softer rocks, and as a result flows parallel to the strike. However, the higher and harder rocks will develop their own underground water levels called water tables and produce springs, and the streams from these will run down the dip of the strata to join the rivers in the soft rock valleys. This produces a rectangular river pattern with tributaries meeting main rivers at right angles – a pattern called trellised drainage pattern.

A river flowing down the dip of the strata inevitably erodes its bed. As the bed cuts deeper into the water table, the source reaches farther up into the hills. Eventually the river may cut right through the ridge of harder rock, and the river in the next valley may change its course to come down the new gorge so formed. This is what is called river capture, and usually leaves a much reduced river trickling out of the original valley.

Any domed structure, such as the sedimentary beds heaved up over a laccolith (pp. 56–57), will produce a radial drainage pattern in which the rivers flow outward, following the dip of the strata.

An example of the radial drainage pattern is the English Lake District. The rivers flow outward from the center, but there are no dipping beds to indicate their origin. In fact, there was once a dome of younger rocks here, on which the rivers started. Eventually all the rocks that formed the dome became eroded away – not just by the rivers

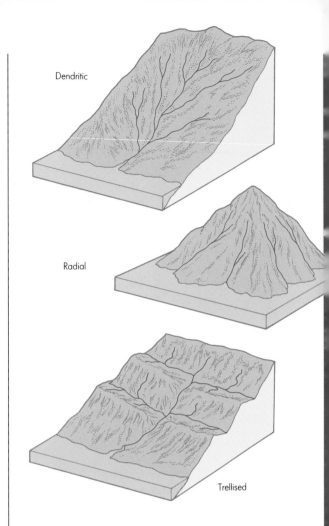

Dendritic

Radial

Trellised

Above The most common river system conforms to the dendritic drainage pattern. Streams flow into one another and form tributaries in a random manner. A radial drainage pattern occurs when streams flow outward from an upland area. A trellised pattern occurs when streams follow the grain of the landscape (the strike) and are connected by main streams running at right angles to them. Occasionally the headwaters of a stream will erode back until it meets another. This second stream will abandon its original channel and combine with the first, flowing down the first valley. This occurrence is called river capture **below**.

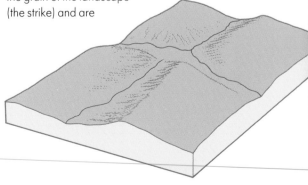

but by the other forces of erosion also. The rivers continued along the courses that they had already established for themselves, irrespective of the new deeper geology over which they were running. This is what is known as a superimposed drainage.

Rivers with a memory

A river, in its mature stage, flows in a meander along its valley floor. If the whole region is uplifted, the river will begin to cut down into its bed. The result is a sinuous gorge following the route of the original river bed and called an incised meander.

Probably the most spectacular example of superimposed drainage is the Brahmaputra. Just look at it on a map: it rises in China on the north flank of the Himalayas, cuts south in a gorge through the highest mountain chain in the world to India, and reaches the Indian Ocean in Bangladesh. It was obviously there before the Himalayas existed, flowing south from the Asian continent. Then when plate tectonics brought India into collision with Asia, it began to cut down through the hills that arose, and then through the mountains that were subsequently thrust up, the rate of erosion keeping pace with the rate of uplift.

The constant shift of river patterns opens up new rock exposures for the practical geologist.

Trellised drainage, as in this Greenland example **top right**, is the result of river capture. A river will cut back through a ridge, forming a valley that will be followed by rivers originally flowing at the other side of the ridge. A river flowing over a flat plain will naturally form loops called meanders. If the flat plain is subsequently elevated by Earth movements the river may cut down into its bed, following the course of the original meander. The result is an incised meander, such as this one in Colorado **right**.

77

UNDERGROUND LANDSCAPES

The rain falling on the landscape mostly percolates into the ground. It gathers in the saturation zone, where all the pores and crevices of the rock and soil are filled with water. The top of this zone is the water table – an important concept in engineering and drilling wells. Where the water table reaches the surface, as on a slope, the water seeps out in the form of a spring.

In a limestone terrain the situation can be more complex, and more spectacular. Limestone is made of calcite (pp. 70–73) which dissolves away in the carbonic acid of the rain water. On the surface this solution takes place most quickly along the joints and fault planes, opening up crevices and separating the limestone mass into clints. This takes place underground too.

Most erosion follows the bedding planes of the limestone and the joints that tend to cut the bedding plane at right angles. It also takes place along the water table, where the surface of the water may form a stream and flow more or less horizontally. As a result a limestone terrain becomes dissolved into a series of interlocking caverns. From time to time the water table may drop. A stream in a tunnel will then erode deeply into its bed and give a tunnel that is keyhole-shaped in cross-section. If the water table drops suddenly (in geological terms) the stream will begin to erode a new tunnel at the new level and leave the old one as a dry gallery. As the caverns expand, their roofs collapse, filling their floors and opening new spaces above. The collapse may cause the surface to cave in, leaving long gorges along the course of underground rivers, or broad depressions called dolines.

Vertical caves carved out by falling water are called potholes (not to be confused with the potholes that are carved out by swirling stones in a youthful river bed) or sink holes. Surface streams flowing off an area of impermeable rock may suddenly disappear down one of these when it meets limestone.

Some calcite is washed away to sea, but much of it is redeposited in the same area. Groundwater seeping through and hanging as a drop on the cavern roof may deposit its calcite there. Not because the water evaporates away – the humidity of a typical cavern would preclude this – but because the carbon dioxide is lost from the water and it ceases to be acidic enough to hold the

Below and far right
Groundwater sinks into holes and cracks dissolved into limestone 1. At the water table it will flow horizontally, dissolving out a tunnel. Around it other openings will be dissolved along natural bedding planes and cracks in the rock 2. If the water table is lowered, the process is repeated at the lower level, leaving the original tunnel as a dry gallery 3. The ceiling may fall, opening up huge caverns.

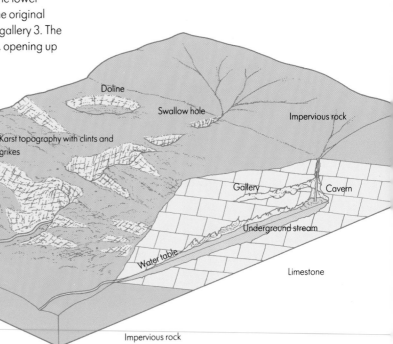

Doline

Swallow hole

Impervious rock

Karst topography with clints and grikes

Gallery

Cavern

Resurgence

Underground stream

Water table

Limestone

Impervious rock

Above A stream flowing onto a limestone surface may dissolve away a vertical shaft for itself, forming a sinkhole. In the underground caverns and hollows, the dissolved calcite is redeposited on the ceilings and floors in the form of stalactites and stalagmites **below right**.

calcite. Accumulation of these calcite deposits builds up stalactites. When a drop hits the cave floor, the calcite is knocked out of it and an accumulation here forms a stalagmite. Different shapes of stalactite and stalagmite develop, each with a descriptive name. Water drawn along a stalagmite or a stalactite by capillary action, for example, will deposit its calcite seemingly at random and produce a twisted stalactite called a helictite.

Calcite is deposited by agitation in the underground stream as well. As an underground stream flows over an irregularity it deposits calcite, which causes a bigger irregularity which deposits more calcite, and so on. The result is a sequence of steps and terraces in the stream bed looking just like a hillside of terraced paddy fields – structures called gours.

When the underground stream finally reaches the surface, it may form a petrifying spring. Here the water can evaporate and deposit its calcite on anything handy. Mosses are sometimes encrusted, as are trinkets left by sightseers.

Calcite is an important mineral in cementing unconsolidated sediment to form a solid rock. A visit to a petrifying spring where the speed of calcite deposition can actually be seen is a memorable demonstration of this.

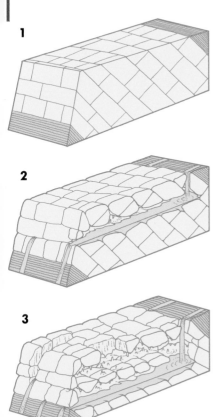

ICE

Ice can be a great rock destroyer, as we have seen (pp. 70–73), but it can also be a great rock mover and landscape sculptor. In tundra regions, fringing the great ice caps of the colder corners of our globe, the permanently frozen subsoil – the permafrost – has all kinds of influences on the landscape, crazing the surface into huge polygons many yards across, or heaving up vast soil-covered, ice-filled blisters called pingos.

Flowing ice

It is the work of glaciers, those rivers of ice that creep down from the snow-capped mountains, that has the greatest influence on our planet's surface. Snow builds up in a mountain hollow, year after year. Eventually the great weight of the top layers compresses the bottom layers into ice. Under great pressure ice can move like putty and the whole mass creeps downhill. As it goes, its great weight grinds out the floor and sides of its valley, and carries the resulting debris along. Valley walls are undercut and avalanches bring more debris down on to the icy surface. A glacier is like a huge conveyor belt for rocks and stones.

More impressive are the ice caps – vast masses of ice that sit over the Earth's poles or on ice-bound continents, and creep outward as more snow falls in the center.

Evidence of ice past

This may not seem to have much relevance to the landscape of the more temperate parts of the world. However, we have just seen the back of the Great Ice Age that has affected the world for the last two million years. Much of North America, Europe, and Asia were covered in ice sheets, and the glaciers extended down from the mountain valleys in other parts of the world. Everywhere in these areas we see landforms that have been sculpted by the ice masses.

Below left and right The great weight of a glacier grinds down the floor and sides of its valley, carrying along the debris it has torn out. When the glacier has eventually melted, the valley will have a characteristic U-shape. The floor may consist of smooth polished rock, and the material carried along – the moraine – will be deposited elsewhere, either as formless sheets of boulder clay, or as hummocks called drumlins. The path of a former stream under the ice may be marked by a ridge called an esker.

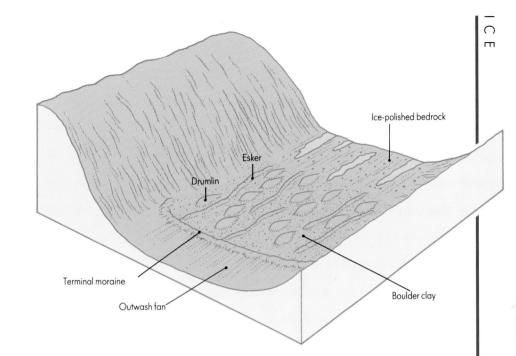

Ice-polished bedrock

Esker

Drumlin

Terminal moraine

Outwash fan

Boulder clay

The deepened and broadened valleys have taken on a distinctive U-shape, with their tributary valleys left hanging halfway up the side walls. When submerged by the sea, these form fjords. Bare rock faces have been gouged with deep scratches called striations. Exposed outcrops have been polished and striated on their upstream side, and shattered and pulled away on the downstream side, leaving the outcrop as a structure called a *roche moutonneé*.

The redistributed debris
Geologists give it the term moraine. It lies in the lowlands where the glaciers finally melted. Huge boulders, coarse rubble, and fine clay particles are all deposited together when the ice melts to form boulder clay. This can be a simple layer over the landscape, or it may form mounds called drumlins several hundred yards or meters long and orientated in the direction of the original ice flow. Sometimes a long winding structure forms, like a sinuous railroad embankment, produced where a stream in a tunnel within the glacier carried rubble along and deposited it at the glacier's snout. This structure is termed an esker. Where the glacier melted back in stages, the moraine lies in curved ridges, each ridge marking the former position of the glacier's snout.

As a moraine contains debris that has been lifted and carried along by the glacier, it can be used to detect the direction from which the glacier traveled. Erratic blocks are particularly useful for this. These are very large boulders that have been shifted by a glacier and dumped somewhere where the geology is entirely different. For example, the erratic blocks found on the east coast of England are composed of rocks found in the Norwegian highlands.

Something to do

Find an outcrop of boulder clay. (Local geological excursion guides will help you find such a place.) Gather a sample of it. Do a rough analysis of the size of particles – dig out boulders and cobbles and put them on one side, put pebbles on another, regard the clay groundmass as the finest material. You will find that they have been all mixed up together. A glacier does not discriminate the size of material it carries. It is strong enough to carry the heaviest, yet it can grind things down to the finest grade. Then, when it melts, it dumps everything haphazardly.

COASTAL EROSION

The sea with its churning currents and pounding waves is, not surprisingly, a very powerful erosional agent.

The physical pressure of walls of water thumping against exposed rock compresses the air in pores and cracks in the rock, forcing it to expand explosively as soon as the pressure is released. This pressure is on average nearly 2500 pounds per square foot (10 tonnes per square meter) but during a storm it can be five times as great. Soft cliffs can be worn back at a rate of one yard or meter per year by this hydraulic action.

Boulders and pebbles picked up from the sea bed are hurled at cliff faces, smashing off more fragments. This process is known as corrasion. The boulders and pebbles themselves are smashed and ground down in the process known as attrition.

There is also a chemical reaction between the salts in the seawater and the minerals of some rocks, and a biological reaction involving grazing seasnails and certain kinds of rock-dissolving bacteria. All these effects combine to erode the sea cliffs.

The crumbling headlands

The erosional effects tend to be concentrated on headlands. Waves attack along a front, defined by the direction of the wind and the shape of the sea bed. As waves approach a headland, they slow down as they meet the shallow water near the headland itself, but continue at the same speed at

In a cliff eroded by the sea, the softer beds will erode first, leaving the harder beds jutting out **top**. When a sea cave forms, the waves funneling into it will build up a tremendous air pressure, forcing the rocks apart. Any crack forced through to the outside will produce a blast of air and water every time a wave crashes into the cave below. This produces a blowhole **left**. Headlands are eroded from both sides. The caves and blowholes eroded along the softer rocks and weaker zones may meet in the middle of the headland and form a tunnel or a natural arch. Eventually the roof of the arch will fall, leaving the seaward portion standing as a seastack **below**.

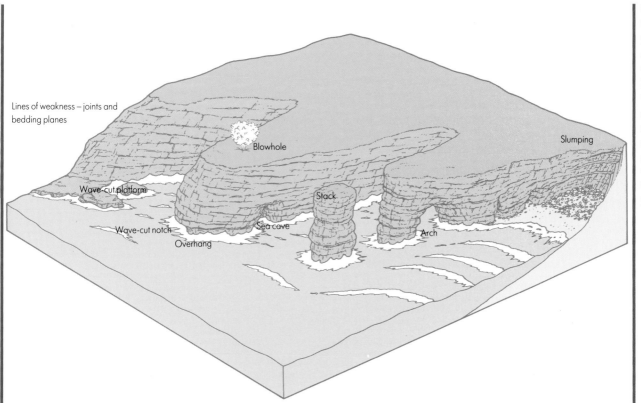

Lines of weakness – joints and bedding planes

Blowhole

Slumping

Wave-cut platform

Stack

Wave-cut notch

Sea cave

Arch

Overhang

each side. As a result the wave front curves around and attacks both flanks of the headland. Cracks are widened into sea caves, and these are themselves widened by the explosive effect of the hydraulic pressure on the trapped air. Occasionally a hole is blown through the roof, and the incoming wave has a piston effect, blasting air and spray out as a blowhole. Caves on opposite sides of a headland can meet each other, producing a tunnel or a natural arch. Eventually, as the arch is worn away, the lintel falls, leaving the offshore part of the headland isolated. The result is a sea stack, which is itself eventually demolished. Coastlines are continually eroded back by such repeated demolition of headlands.

Wave erosion is confined to the areas that waves can reach. When a sloping coastline is attacked, the waves erode the few yards or meters close to sea level. As the coast is worn back, this cuts a notch which can be quite an even and graceful curve in homogeneous rock. The resulting overhang eventually collapses into the sea. The debris is soon cleared away by current and wave action and the undercutting continues. Erosion takes place by the retreat of the cliff so

formed. The shape of the cliff is determined largely by its geology. Soft material such as sand or clay forms a sloping crumbling cliff, while something with a relatively even composition and texture forms a fairly smooth vertical surface. Strongly bedded or deeply jointed rocks erode irregularly, erosion being swiftest along the planes of weakness.

Coastal patterns
On a wider scale, the regional geology determines the nature of the coastline. Strata that run parallel to the sea produce a so-called Pacific type coastline with inlets that cut through the ridges and spread out in broad inlets along the softer rock – as in San Francisco Bay – and elongated islands and reefs parallel to the coast. On the other hand, strata that run straight out to sea produce a deeply indented coastline, with the hard rocks forming long headlands and the soft eroding to long bays. Southwestern Ireland is a magnificent example, and the result is called an Atlantic type coastline. The constant opening up of fresh cliff faces along an area of coastal erosion makes such places ideal prospecting sites.

COASTAL ACCRETION

The waves can destroy. They can also build. The boulders, cobbles, and pebbles washed off the headlands and the cliffs, and the sands and gravels ground down from the bigger fragments, are all washed about and redistributed, eventually ending up deposited someplace where the land is being built up rather than destroyed. Such places are seen as beaches, sand spits, and sand bars.

Beaches on the march

Two important concepts in this are longshore drift and beach drift. The first is caused by currents that sweep along the coastline, moving material parallel to the shore. The second is caused by the waves that usually strike the shore at an angle. As they do so, any pebbles or sand grains that they carry are washed diagonally up the beach. As the water recedes after every wave it pours straight back down the slope of the beach, carrying the debris with it. The next surge of the wave carries the fragments diagonally up once more, and then straight back down. As a result, each piece of beach material follows a zigzag route along the beach. Coastal communities often fear that this action will take away their beaches and so barriers are installed, projecting into the sea at right angles of the coast. These are called groynes, and trap the beach sediment on one side. From the air the resulting beach has a distinctive saw-toothed pattern.

Bars and spits

Beach drifting carries the sand along the coastline until it reaches an opening or a river mouth. There it continues in its journey and builds out a tongue of sand called a spit. The wave front, like that at a headland, curves around the end of the spit and deposits more sediment on the back of the tip. As a result, a typical sand spit has a hooked end which appears to curve upstream. Such a spit rarely goes all the way across a river mouth, since the current tends to carry much of the sediment away and leave an opening. However, the presence and the continual buildup of a spit may move a river mouth along the coast in the direction of the prevailing wave fronts, so that the river now reaches the sea many miles or kilometers away from its original mouth. A good example of this is the River Alde in eastern England. A sand spit called Orford Ness, built up by the currents and waves of the North Sea, has moved its mouth about 5 miles (8 km) south of its original site.

When a spit is built up right across the mouth of an inlet the result is a sand bar, cutting off an area of water to form a lagoon. Many of these are seen around the south coast of the Baltic Sea. The region of the Baltic is slowly rising, and so the sea is constantly changing its shape, and the deposition pattern is changing with it. A sand spit often reaches out to an island, linking it to the mainland. The resulting landscape feature is called a tombolo – an Italian name owing to the fact that many of them exist along the west coast of Italy.

Left A combination of sea currents and waves carries sand along a shoreline. To prevent beaches from being washed away, the authorities often build barriers, groynes, running out to sea. The sand builds up on the upstream side, but continues to be washed away from the downstream side, producing a zigzag beach pattern.

Right Waves usually approach a beach at an angle. Sand particles are washed up the beach at that angle. When the wave recedes, the water and the sand particles wash directly down the slope. The next wave washes them up at an angle again. As a result, sand moves along a beach with the constant wash of the waves.

Left The moving sand will build out beyond the end of its beach as a sand spit. If this spit continues right across an opening, cutting off a lagoon, it is called a sand bar. Sometimes a bar will connect an island to the mainland, such as here in the Indian Ocean. The resulting feature is called a tombolo.

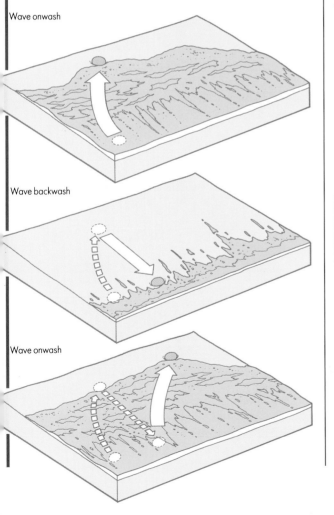

Wave onwash

Wave backwash

Wave onwash

A conflict of currents
Beach drift moving in from two different directions may bring sediment that piles up in a low triangular headland consisting largely of gravel ridges. One of the best examples is Cape Canaveral in Florida, where gravel has been piled up by conflicting eddies from the Gulf Stream. Sand and other light materials brought up by the waves may then be picked up and carried along by the wind. This is the origin of the sand dunes that often back up exposed ocean beaches.

Something to do

Go to the beach in the summer. Take a photograph of the beach from a particular point. Take samples of the beach material – sand, gravel, or whatever – at that time. Go back to the same spot in the winter. Take another photograph from exactly the same position, and gather another sample of beach material. Then do the same thing the following summer.

The changes in beach material will show the different energies of waves – coarse gravel is shifted by much fiercer waves than fine sand – and the photographs will show marked changes brought about in a very short time.

85

HOMO DESTRUCTUS

The most potent agent of geological erosion is, undoubtedly, ourselves.

This can sometimes be observed while watching a geological field party from a college or a natural history society bashing away at a significant rock face, reducing it to rubble in order to extract some pretty crystals. Even worse are professional fossil collectors who will bring along tractors and bulldozers to remove the overburden from the bed in which they are interested.

The effect of civilization

Yet even this is small in scale compared to the work of civil engineers. To build a dam or a bridge, they blast away mountainsides to find firm footings for the foundations, and quarry deep pits for the raw materials. The biggest hole in the ground made by people is Bingham Canyon copper mine in Canada reaching a depth of 846 yd (774 m) over an area of 2¾ sq miles (7.1 km^2). Shallower excavations in the form of strip mines can turn over the top layer of rocks for vast areas. Farmers, too, have a profound effect on the landscape, and have done since farming was developed in Neolithic times. Natural vegetation is removed to make room for cultivated land. Forests are cleared to make grasslands for grazing animals. If this is mismanaged, as it is in many parts of the world, the soil is destroyed. The naturally formed humus that holds the soil together deteriorates and the soil washes away as mud or blows away in sandstorms.

A scheme to irrigate the desert lands of central Asia by diverting the rivers flowing into the Aral Sea backfired when the sea began to dry out and the exposed salt was blown over the new farm-land, making it unfit for cultivation.

Coastal erosion is affected as well. When beach gravel is removed for building material, the pattern of coastal currents is changed and nearby villages may collapse into the sea. Breakwaters constructed to protect harbors may give rise to sandbanks where they are neither expected nor desired.

Atmospheric changes

It is not just the mechanical aspects of weathering and erosion that are affected by human activity. The pattern of chemical weathering changes as

The sheer scale of human activity is perhaps the biggest problem of geological erosion. The conical piles **top** are spoil dumps for china clay pits. **Right** The massive Bingham Canyon copper mine. Much damage is done unwittingly, **above**, where the grasses, eroded by many footsteps, no longer hold the topsoil in place, which means that it is washed or blown away.

well. Smoke from power stations tends to be rich in sulfur. This combines with moisture in the atmosphere to form sulfuric acid. This will later fall as acid rain, affecting buildings made of limestone. It also kills vegetation, with the knock-on effect that this has on soil erosion.

Carbon dioxide, water vapor, and other gases pumped into the atmosphere as a side effect of industry change the balance of atmospheric gases and produce what is known as the greenhouse effect – heat from the sun can pass through to

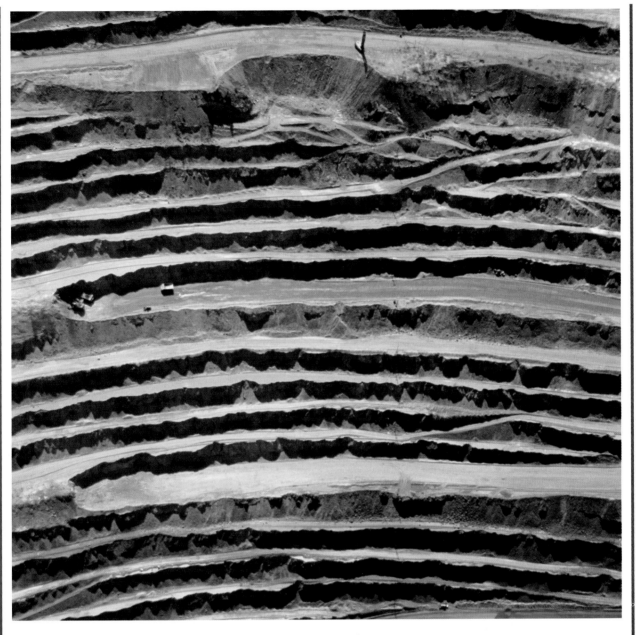

Earth's surface all right, but the re-radiated heat finds it difficult to escape. As a result, the climate seems to be changing, and may change even more drastically in the future, bringing a shift in the climatic zones of the world and a change in the natural vegetation.

The climatic changes brought about by the Ice Age during the past two million years are well documented in the geological record. Imagine how the Age of Humans will show up to whatever geologists will be studying it in tens of millions of years' time. The modern age will be reduced to a bed an inch or two, or a few centimeters, thick, rich in exotic minerals, stained red with iron and green with copper.

And it will probably mark a mass extinction – a phase in the fossil record where a large number of species become extinct to be replaced by something completely different, or even nothing at all. Sobering thought, isn't it?

FIELD WORK TECHNIQUE

*As well as knowing what
to look for, it is essential
to know how to go
about finding it,
and recognizing it
when we see it.*

TECHNIQUES FOR IGNEOUS ROCKS

In a region of igneous rocks, it is a good idea to look first for the the contact between the igneous rock and the original rock. Finding it will be a great help to the mapping of the area. If the boundary between the igneous and the native rocks is well defined, note its strike and dip, determine whether it cuts across the "grain" of the original rocks or is concordant with it, note whether it is a sharp boundary or diffuse, see what effect it has on the wall rocks, and observe any metamorphism that has taken place.

Structures and shapes

From these observations you should be able to determine whether the rock is in the structure of a batholith, a dike or a sill, or a lava flow.

Right In a landscape of extrusive igneous rocks, the soil underfoot is formed from decaying lava – either basaltic or andesitic. Old lava flows may be obvious in the outcropping rocks. The horizon will consist of the conical peaks of old volcanoes. A landscape of intrusive igneous rocks, however, may be similar to that produced by metamorphic rocks.

Now look for any cooling features. These would include the columnar jointing we would expect to see in a quickly-cooled basalt flow, or a chilled margin in which the mineral crystals close to the contact with the native rock are more finely grained than the rest of the mass, or extensive veins that filled with minerals at a late stage of a slow cooling.

Crystals and inclusions

Look for vesicles – bubbles of gas like those trapped near the top of a basaltic lava flow. Look for amygdales – exactly the same but filled with mineral. This would help to distinguish whether a lava flow was recent, with vesicles, or old, in which case there would have been time for the vesicles to be filled with minerals deposited by groundwater. Determine the mineral content of the amygdales – usually calcite or quartz.

See if there are any flow structures within the

igneous mass, such as concentrations of crystals toward the center of the structure. Note any signs of layering, in which some minerals may have solidified and settled while the bulk of the mass was still liquid. Note the presence of phenocrysts – crystals that are much larger than those in the rest of the mass. These would suggest that the igneous mass began to cool slowly and then finished cooling much more swiftly.

Look for xenoliths – chunks of the original rock that were caught up in liquid flow and have become embedded in the final igneous rock. These will be metamorphosed. Note the effects of metamorphism, both on the xenoliths and on the native rock round about.

Note if there is only one phase of igneous activity, or if there are several – with dikes cutting across sills. Note the orientation of all structural features, such as joints, and try to discern a pattern later.

What is it made of?

When you take a sample, number it and enter the observation in your notebook or map. On the spot you can note the size and shape of the specimen – important if the specimen is a volcanic bomb or a xenolith – the color and density, giving a rough guide to the mineralogy, and the texture and grain size. Identify as many of the constituent minerals as possible, using the hand lens and your other field kit such as your hardness points and your streak plate. You should have enough information by now to give the rock sample a provisional name. Then pack it up safely so that you can study it at leisure when you get back home or to the laboratory.

TECHNIQUES FOR SEDIMENTARY ROCKS

Take a walk up a stream bed. Gather up the pebbles that interest you. Chip away at a fossil exposed in the bank. Pry a well-formed crystal from a vein. You are now a rockhound! You are not being a geologist.

Patient and methodical field work is needed if your experience and findings are to have a lasting value and enjoyment. Develop a routine.

Before beginning your trip, take a note of the

Right A typical sedimentary rock is made of layers, or beds, and these may be visible as the rock outcrops. Usually the beds are sloping, or dipping, and this gives rise to a "scarp and dip" topography. A gentle slope formed by the surface of the beds is the dip, while the steep rough slope, where the beds are broken away, is the scarp.

date in your field notebook, and make some general remarks about the nature of the work you are about to undertake. By reading your map and looking at the area from a distance you will see the most likely places for outcrops – disused quarries, stream gullies, highway excavations, and so on. You will be able to work out a makeshift itinerary.

At each outcrop, take a note of its position and mark it on your map. If you are in open country, far from any mapped feature, take cross-bearings from recognizable landmarks or from your starting point. Note the kind of outcrop – an isolated exposure, a cliff face, a stream section, or what-ever. If it is interesting enough, sketch it in your notebook, recording the scale of the sketch and the direction from which it is viewed.

Measurement of the beds

In a sedimentary rock, measure the dip and the strike by using the clinometer and a compass. Enter the readings on your map, using the conventional symbols. Beware of the apparent dip. It may be that an exposure in a vertical sea cliff face shows a bed to be dipping, but the orientation of the cliff face may not be parallel to the dip, so the dip reading will be wrong. You must find a three-dimensional exposure, possibly in the wave-cut platform at the base of the cliff, in

order to find the strike and the direction of true dip. If you cannot do this, then the apparent dip should be noted and marked as such. The dip may be irregular, as on a surface showing current bedding or ripple marks, in which case you must make several measurements over as large an area as possible, and take an average.

Note also the bed's thickness. Always measure the thickness at right angles to the dip.

The structures

If the outcrop is of a complex nature – as in two sequences of rocks separated by an unconformity – study it systematically in its logical sequence. Describe the lower beds first. Then deal with the nature of the unconformity, noting whether or not the first series of beds is eroded irregularly with a conglomerate on top (as would have happened when the sea flooded the land) or planed flat (as would be formed at the bottom of the sea). Then study the overlying beds.

In any sequence, study the rocks in their order of deposition, starting at the bottom and working upward. Make a rough identification of the fossils you find, selecting the more interesting ones for careful examination later.

Note the post-depositional structures – the folding and the faulting. Record the orientations of fold axes, faults, and joints. Do as many as you can and try to work out a pattern for them later.

Every sample you take must be labeled immediately. Attach a gummed label with a serial number that you note in your field notebook or on your map. Wrap it carefully and pack it for safe transportation back.

TECHNIQUES FOR METAMORPHIC ROCKS AND DRIFT

Thermal metamorphic rocks are usually studied at the same time as the igneous intrusion that metamorphosed them. Regional metamorphic rocks, on the other hand, tend to be the subject of separate expeditions and studies.

Looking at regional structure

At each outcrop observe the texture, the banding, the foliation, and the schistosity. Note the orienta-

Which way up?

Sometimes in highly contorted terrain it is difficult to see which way up the beds should be. The concept of younging comes into play. Look for features that could only have been deposited one way up – graded bedding has the coarsest material at the bottom, current bedding is concave upward, as are sole marks and tool marks.

tion and inclination of these. If the original rock has been sedimentary, see if there is any sign of the original bedding, and note the dip and strike of this. See also if there are the remains of any fossils or other sedimentary features, and note the extent of their deformation.

Regional metamorphic rocks tend to occur in mountainous areas, and so such an expedition may take you to remote places. You should bear this in mind when collecting specimens. A small specimen should be good enough for a fine-grained rock, but if you get back and find the small specimen inadequate it will probably be a long hike back to get another one. As usual, trim your specimen to show fresh faces, but if the weathered face helps to show up the texture then preserve this as well.

When noting the texture and structure of the rock, see if the crystals are of the same size, or if some are larger than the rest.

The loose stuff

You are unlikely to find that the rocks outcrop over the whole of your area of study. Most of the area will be covered with soil, scree slopes, flood deposits, and other natural material. This is referred to as drift on geological maps, as opposed to the underlying geology which is termed solid. You should become familiar with the features of drift, and how they relate to the solid rocks beneath. Look at the kinds of stones thrown up in rabbit holes and other workings, note any change of vegetation, see if a slope alters its angle – possibly indicating a bed of harder rock.

The study of drift is itself interesting. You will soon be able to distinguish the drift that forms *in situ* from that which has been deposited by other means. If the loose material contains fragments of the underlying rock, then it is safe to assume that it it is a soil formed from the decay of the rocks in that place. Boulder clay, on the other hand, is a jumbled mixture of all sorts of rocks brought in from other areas by ice. This can be something of a headache since it may be dozens of yards or meters thick and effectively isolate the geologist from any solid native rock.

Above Metamorphic rocks, particularly regional metamorphic rocks, tend to be found in the deep interiors of mountain chains, since they are formed by the mountain-building process. In very ancient shield areas, however, the Precambrian mountains have been worn flat, and the result is a low rounded topography.

Blasted geology!

In igneous rocks a texture in which large crystals are scattered through a fine groundmass is called porphyritic. In metamorphic rocks it is called *porphyroblastic*. The prefix "blast-" is one that geologists often use in the study of metamorphic rocks. It means the metamorphosed remains of something else. Hence we get poikiloblastic – a texture in which a fine mass of new minerals forms around the original minerals, and blastopsammite – a piece of sandstone embedded in a metamorphosed conglomerate. Porphyroblastic, however, refers to the new crystals that have grown with the metamorphism, and should not be confused with *blastoporphoritic* in which we see the remains of a porphoritic structure that was present in the original rock.

Confused? Don't worry. Familiarity with specialist terms comes with experience.

GRANITE

The proportions of the chemical components of the continental crust are such that, if they were to be melted down, mixed thoroughly, and slowly cooled, the result would be granite.

Mineralogy

It consists largely of feldspar, quartz, mica, and a small quantity of iron ore.

Landscape and structures

Granite landscapes are usually hills or moorlands. Granite forms in batholiths, deep within mountain chains, and so granite landscapes usually occur where very ancient mountains have been eroded away, for example in the Appalachians or the western peninsula of England. Outcrops also occur where rivers have eroded deeply into mountainous areas, as in the Grand Canyon.

In humid climates the feldspar in the granite tends to decay along joints and cracks, breaking any outcrop into rounded blocks and producing castle-like structures called tors, particularly on moors and along coasts. Take a note of the orientation of these cracks. They may give a clue to the pressures that have affected the region since the granite was emplaced. The feldspar decomposes to produce china clay (kaolin), and so granite moors are often characterized by the white spoil-heaps from china-clay quarries. The decayed feldspar loosens the quartz and mica from the rock and this is washed away, producing dazzlingly white beaches by granite landmasses. Drainage is poor through granite, so bogs tend to form. Take waterproof footgear when looking at granite terrain.

In arid climates most weathering is on the surface rather than along joints. Onion-skin weathering and sand-blasting produce rounded hills called inselbergs or kopjes, such as those that dot the Serengeti wildlife park in Tanzania.

If you are lucky you may find inclusions – chunks of the surrounding rock that were engulfed by the granite when it formed. These will be metamorphosed beyond recognition. You may even see the contact between the batholith and the country rock, and find more metamorphism. The country rock will probably be full of veins, as may the edge of the granite itself, where hot fluids filled cracks as the igneous mass finally cooled. These veins will be made of large crystals of quartz, and may contain large crystals of other minerals as well. The extremely coarse rocks formed in veins are often called pegmatites.

Right Because of the feldspar content, granites are subject to chemical weathering in moist climates. Joints are eroded and the intervening masses are left as rectangular blocks reminiscent of massive masonry. This example is at England's Land's End.

Hand specimen The surface of a granite outcrop is invariably weathered, and so it is necessary to break it open with a hammer to see a fresh face. It is a good idea, however, to retain a weathered face – it may help to show up the texture. There is no internal structure or planes of weakness, and so it will break irregularly. The grain is usually so coarse that you can see the individual mineral crystals –

the glassy quartz, the milky feldspar, and the dark but sparkling mica. The iron ore will be too small to see.

Microscopic specimen Through the crossed polars of the microscope the most obvious crystals are those of the feldspar. They will usually be twinned, each half of the twin showing a different extinction angle and fading from white to black through shades of gray (see pp. 28–29). The shapeless pieces of quartz

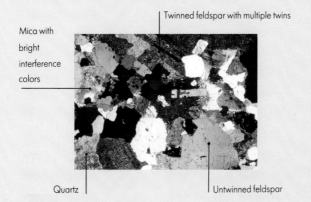

Mica with bright interference colors

Twinned feldspar with multiple twins

Quartz

Untwinned feldspar

are somewhat dull and featureless. The mica will show a range of colors and will have irregular shapes. The iron ore magnetite will be present as tiny grains and will be seen best through a single polarized filter (the polarizer) as opaque fragments amid the general transparency.

Varieties The chemical nature of the feldspars will alter the appearance of a granite outcrop. If the feldspar is rich in potassium it tends to be white in color. The white feldspar and quartz contrast with the black of the mica to give a speckled appearance and an overall gray color. Sodium-rich and calcium-rich feldspars are pink and impart an overall pink color to the rock.

A porphyritic texture is sometimes in evidence, in which the feldspar crystals are so much larger than the others. The very large crystals in a porphyritic texture are called phenocrysts.

The large crystals give granite an attractive appearance, which is why it is often used for building and

Coarse porphyritic granite

ornaments.

Similar rocks Syenite is fairly similar in its occurrence and appearance although much less abundant. It is a coarse-grained intrusive rock of intermediate composition – having no quartz but some of the magnesium-iron silicates such as amphibole and pyroxene. The resulting rock has a slightly darker color. The coarse metamorphic rock gneiss has a similar grain size and mineralogy to granite, but it can usually be distinguished by the fact that the minerals form in discrete bands (pp. 116–117).

DOLERITE

Dolerite is a medium-grained igneous rock and is found in small-scale and medium-scale igneous structures, such as sills, dikes, laccoliths, and volcanic necks (pp. 56–57).

Mineralogy
Dolerite is a basic rock consisting of olivine and pyroxene, with some feldspar and mica. Apatite is usually present as well.

Landscape and structures
The small-scale igneous structures in which dolerite forms may be harder or softer than the native rocks into which the magma was injected. Usually the dolerite is harder than the surrounding rock and the structure stands proud as the surrounding rocks are eroded away. A dike will form a vertical wall across the landscape, often very visible at a site of coastal erosion. Where the native rock is more durable the dike will erode more quickly forming a trough. A sill will give the appearance of a particularly robust bed among the other sedimentary beds, and will show as a cliff or a scarp slope. A dolerite sill may be hundreds of yards or meters thick and form a flat-topped mountain. Thermal metamorphic rocks will be found both above and below.

The magnesium-iron minerals in it, particularly olivine, are subject to chemical alteration by the weather, and an exposed dolerite may erode quickly into spherical shapes with flaky surfaces.

Always note the orientation of a dike, and note whether or not it is one of a swarm.

Above When dolerite occurs as a sill it is usually harder than the rocks round about. It tends to protrude from the landscape as a prominent cliff, following the grain of the land, as along the rim of this canyon in Wyoming. Columnar jointing is also visible here, produced by joints that formed at right angles to the boundary of the sill as it cooled.

Hand specimen Because dolerite weathers so easily, it is essential to break it open to find a fresh face. It is a dark heavy rock with few light-colored minerals. Its dark color may have a greenish tinge due to the presence of serpentine produced by the breakdown of olivine. It has no internal structure and so it breaks into uneven pieces, usually with flat faces and sharp angles. Being a medium-grained rock, it will be difficult or impossible to discern the crystals with the naked eye, and a hand lens will be needed.

inches

centimeters

Above A fresh face of dolerite.

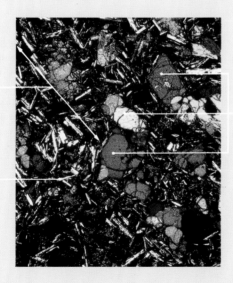

General lineation of feldspars

Feldspar

Olivine crystals
Serpentine in cracks

Microscopic specimen
Through crossed polars the magnesium-iron minerals show up in deep reds and greens. The gray feldspar crystals tend to be long and thin and embedded in pyroxene crystals, giving the so-called ophitic texture. A sample taken from the edge of the intrusion will show the more rapid cooling by the finer crystals, and may have the elongated crystals of feldspar lined up in the direction of flow.

Varieties The proportions of the magnesium-iron minerals vary a great deal. These variations can only be detected in the microscopic section.

Similar rocks The intermediate equivalent of dolerite is diorite. This has no olivine and a much greater proportion of feldspar, which is of the sodium and calcium type. Having no olivine, it tends not to weather so quickly, and it has a lighter color. It is less common as the constituent of small-scale igneous intrusions.

BASALT

Basalt is an extrusive basic igneous rock, forming widespread lava flows, particularly in Iceland, Hawaii, and the Galapagos Islands. Ancient basalt structures are known from all over the world. Much of India is covered by the Deccan Plateau – a series of basalt lava flows some 50 million years old, covering an area of 250000 sq miles (650000 km^2) to a thickness of 7000 ft (2 km).

Mineralogy

Essentially the mineralogy is the same as for dolerite – just more fine-grained.

Landscape and structures

Basaltic lava gushes from a fissure eruption, or from a broad flat shield volcano. As it flows it cools on the surface, forming a leathery skin which wrinkles up with the continued motion of the molten material below. When this finally solidifies, it forms ropy lava, scientifically known by its Hawaiian name "pahoehoe." When the flow is turbulent, the leathery surface breaks down into jagged masses producing blocky lava or, in Hawaiian, "aa." Gases may gather and burst up through the pahoehoe crust forming funnels, waist-high, called hornitos. Tree trunks engulfed by a lava flow will burn to ash, but the cooling lava that encrusted them will be left upstanding as tree molds. Liquid lava may flow away, leaving a solidified outer crust as a lava tunnel, often big enough to walk through.

Always wear good strong boots when looking at basalt terrain. Broken pieces of pahoehoe and aa are sharp, glassy, and splintery.

Old eroded basalts may show columnar jointing (pp. 54–55). The upper surface will be eroded away to form soil which may contain spherical chunks of decaying basalt – since basalt has a similar mineralogy to dolerite it may decay in the same spheroidal manner. Check the soil layers between successive lava flows. You may find fossil leaves from the vegetation that grew on top of one flow before being engulfed by another.

Basalt erupted at the bottom of the sea will quickly cool and form a flexible skin on the surface but remain molten inside. The molten mass will break up into individual clumps that will then roll about on the sea floor until coming to rest and solidifying draped over one another. Such structures are called pillow lavas. Ancient pillow lavas, when cut across, show a concentric pattern, indicating that they cooled from exterior to interior.

Basalt may weather to a whitish patina in arid areas, and to a reddish color in humid areas because of breakdown of the iron minerals.

Left Basaltic lava can cover large areas. When it engulfs a tree, the tree is destroyed, but the lava next to the tree trunk is cooled quickly. The rest of the lava may flow on, leaving a vertical tube of basalt with burned wood in the middle – a tree mold. These examples were formed as a forest was overwhelmed in Hawaii.

Hand specimen A typical hand specimen of basalt will have a ropy texture on the outside, but will be cindery, brittle, and full of bubbles on the inside. Handle pahoehoe and aa specimens with gloves because of the brittle glassy shards. Any hand specimen of basalt may contain bubbles where gas was given off during eruption, and the gases failed to escape before the lava solidified. In specimens of ancient basalt these bubbles may be filled with a mineral such as calcite, deposited by groundwater over millions of years.

Another good hand specimen is the volcanic bomb. This is a chunk of lava that blasted from the vent during the eruption and solidified in the air before landing. Often these are teardrop-shaped, ribbon-shaped, spindle-shaped, or simply rounded rocks, depending on their passage through the air as they

Section through pahoehoe lava

A volcanic bomb

cooled. They can be as big as a car, or small and pebble-sized – in which case they are termed lapilli.

Solid basalt is heavy and black, and the mineral crystals are too fine to be seen by the naked eye, or even with a hand lens. There

may, however, be largish lighter-colored minerals scattered throughout.

Varieties The various types of basalt are classified by the different types of magnesium-iron minerals, and different types of

feldspar, as revealed by careful microscopic analysis.

Similar rocks Basalts are similar to dolerites in composition, varying only in the texture and mode of emplacement.

Microscopic specimen Under the microscope the basalt appears to be a shapeless mass of tiny feldspar crystals. There may be well-shaped crystals of other silicates, such as pyroxene, and these may be much larger than the feldspars. These are due to the early crystallization of the pyroxene before the eruption. Ophitic texture may be present.

Fine groundmass of feldspar showing flow texture

Big crystals of olivine formed earlier

RHYOLITE AND ANDESITE

Rhyolite is the extrusive acid igneous rock, and andesite the intermediate equivalent. They issue as lava from volcanoes in mountainous areas and volcanic island arcs associated with the subduction action of the tectonic plates.

Mineralogy
Rhyolite is the extrusive form of granite, and the chemical composition and mineralogy reflect this. Andesite has less silica and so contains no quartz.

Landscape and structures
Acidic and intermediate lava flows are much more viscous than basaltic ones. As a result, rhyolite and andesite are restricted to areas fairly close to their parent volcanoes. These volcanoes tend to be steep-sided and conical. An active volcano of this type is very dangerous and it is not advisable for the amateur geologist to study it too closely.

The eruptions tend to be explosive and send clouds of ash over large areas. Although the lava itself may not travel far, the ash and dust may settle over many thousands of square miles or kilometers round about. This ash can be light-colored with a texture like cookies, being full of gas bubbles. It may be so lightweight that it can float on water, in which case it forms pumice.

Occasionally there may be acidic and intermediate eruptions in basaltic areas. This is due to a process called fractionation. In this, many of the magnesium-iron minerals have already crystallized out of the magma when it was still underground, leaving a silica-rich liquid that is then erupted at the surface. A large area of basaltic Iceland is covered by a carpet of yellowish ash erupted from the volcano of Askja in 1875. The landscape of black basaltic rocks protruding through this light-colored surface is quite distinctive.

An eruption involving these lavas produces a characteristic tinkling sound, caused by the masses of freshly solidified glassy fragments rolling down the steep flank of the volcano. Sometimes the heavy metallic nature of the sound recalls the shunting of railroad cars.

A fall of andesitic ash in predominantly basaltic Iceland **left** shows the color contrast between the dark basic igneous rocks and the lighter acidic material. Andesitic volcanoes tend to be conical with steep sides, since andesitic lava is sticky and does not flow far.

Eruptions can be explosive and catastrophic.

Above A large breadcrust bomb ejected by a volcano in the Italian Aeolian Isles.

Hand specimen Rhyolites and andesites tend to be fairly light in color and light in weight due to the relative absence of magnesium-iron minerals. They can have a somewhat flinty appearance, splintering when they are struck with a hammer. Often they have a banded texture showing the very viscous nature of the liquid lava. Andesites may have a porphyritic texture, with large crystals of feldspar big enough to be seen with the naked eye.

Acidic and intermediate bombs tend not to be as streamlined as basaltic ones. As the bomb flies through the air the surface may solidify and cool rapidly while the interior is still hot and molten. This gives rise to a breadcrust bomb which has a cracked outer layer. Hot pieces of ash may weld together as they fall, forming lumpy masses called agglomerate. When large

Rhyolite

Andesite

masses of this are welded together to give a continuous rock it is called tuff.

Similar rocks Rhyolites and andesites are fairly similar to one another, differing only in the presence of quartz in the former. They are markedly different from the extrusive basic rock basalt.

inches

centimeters

Microscopic specimen The microscopic view shows masses of tiny feldspar crystals. The lava cools so quickly that crystals may not have time to form, and the whole mass is an amorphous glass. Amphiboles and pyroxenes may make good crystals, having formed before the eruption.

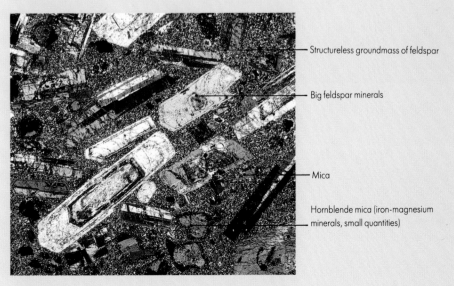

Structureless groundmass of feldspar

Big feldspar minerals

Mica

Hornblende mica (iron-magnesium minerals, small quantities)

LIMESTONE

Limestone is a sedimentary rock, usually biogenic or chemical in origin. It forms in the shallow waters of shelf seas, and so it tends to cover very large areas.

Mineralogy

Limestone is, basically, a rock made of calcite. However, being a sedimentary rock, it may be mixed in with other minerals depending on how clear the water was when it formed. Dolomite – magnesium carbonate – may be mixed in with it, and in some cases may form the major proportion. Sandy and shaly limestones are common, and it becomes difficult to classify the various grades. As a rule, any rock with more than 50 percent carbonate minerals is classified as a limestone.

Landscape and structures

A limestone terrain is dry. When the limestone is massive – when it comes in thick beds rather than thin slabs interbedded with other rocks – it forms arid upland plateaus. Streams vanish from the surface, dissolving limestone and forming underground caverns. The weathering features called clints and grikes (pp. 70–73) are common, as are collapse features such as gorges and dolines (pp. 78–79).

Ancient limestone landscapes may have been buried by later sediments, and the erosional features will be fossilized. The rubble from collapsed cave roofs will be cemented together by redistributed calcite and preserved as collapse breccia. Sometimes this preserves the skeletons of animals that lived there when the caves existed. Carboniferous limestone in southwestern England was exposed in Triassic times and its crevices contain the fossils of Triassic reptiles. The same Carboniferous limestone in Belgium formed a ridge in Cretaceous times and Cretaceous dinosaur skeletons have been found in a gorge there.

Where underground streams reach the surface again there may be deposits of tufa.

An outcrop of limestone in an industrial area may be marked by a line of cement factories.

The rugged waterless nature of a limestone landscape is well-known and is due to the chemical weathering of the chief component calcite. Clints and grikes form on flat-lying limestone beds **above** forming a limestone pavement and giving a topography known as a "karst" landscape, named for the area in Yugoslavia. Rivers cut deep steep-sided gorges, by vertical erosion or by collapse of caverns and underground waterways. The Cévennes region of central France **right** has fine examples, such as the Gorges du Tarn and the Gorge de la Jonte.

Oolitic limestone

Hand specimen

Depending on the type of limestone, the hand specimen may be a mass of fossils or it may be an even-grained, pale-colored rock.

When a biological limestone is weathered, it may show the constituent fossils in high relief.

A variety of the chemical limestone called oolite consists of tiny spherical particles about 1/16 in (1 mm) across, called ooliths. In a coarser variety, pisolite, the grains are pea-sized. These have formed as calcite precipitated on fragments of sand or shell, and then rolled about on the sea floor, building up like snowballs.

In the hand specimen you can distinguish between calcite and dolomite by using acid (vinegar will do). Calcite fizzes and bubbles, but dolomite does not react.

Varieties There are many types of limestone. The shelly limestones are usually classified by the kinds of fossils they contain, for example gastropod limestone, crinoidal

inches

centimeters

Shelly limestone

limestone, or coral limestone. Many of these take a good polish, revealing light fossils contrasting with a dark matrix, so are used in building. They are often misleadingly called marbles by stoneworkers; for example, Purbeck marble (a Jurassic limestone with gastropods) and Forsterly marble (a Carboniferous limestone with corals).

Limestone conglomerate consists of small pieces of preexisting limestone, called by the mineralogists intraclasts, cemented together.

There is also a type of limestone that is made up of calcite pellets – similar to ooliths but with no internal structure.

The nature of the cementing groundmass is important. Those cemented together with coarse crystals are called sparites, and those with very fine crystals micrites.

Microscopic specimen

Fragments of the fossil content are usually very recognizable, with the cementing calcite visible as a regular mosaic around them. Sometimes the cementing calcite is grown from an existing calcite fragment and you can see the shape of the original as a ghost in a larger calcite crystal.

In an oolite, the concentric shapes of the individual ooliths are very visible, again cemented together by a calcite mosaic.

Photomicrograph of shelly limestone

CHALK

Chalk is a particularly pure type of biogenic limestone. It formed in extensive massive beds on the floors of broad shallow continental seas at the end of the Cretaceous period.

Mineralogy

Chalk consists almost entirely of calcite, in one form or another. It can be 98 percent pure. It consists of the calcite shells of microscopic marine algae which may or may not be filled with crystaline calcite. As a rule the chalk contains no land-derived sediments whatever, but older beds of chalk may contain some mud particles, and when these become so great that the purity of the calcite drops to about 80 percent the rock becomes chalk marl.

Landscape and structures

A chalk landscape is quite distinctive, forming the rolling downland that is so characteristic of southeast England and northeastern France.

Chalk usually forms grassland as trees do not grow well on it. Beeches are the main trees found, and these usually form discrete thickets and woods. As with other types of limestone terrain, there are few surface streams. However, broad, gently undulating dry valleys abound; these may have been formed during the Ice Age when the groundwater was frozen.

At the sea it forms spectacular white cliffs, such as the famous White Cliffs of Dover. The regular texture means that the cliffs are vertical and straight, and tend to be eroded back evenly, producing a straight coastline.

Silica, which may have been present in the original sediments, collects in distinct levels, as irregular bands of flint lumps, or as beds of chert.

The startlingly white appearance of the naked rock contrasts strongly with any vegetation cover. Pits from which chalk has been excavated are visible from a long way off. Often this is exploited as an art form and shapes such as horses and giants are cut into chalk hillsides by removal of the turf.

Chalk downland is waterless, rainwater seeping through the porous rock to form streams. Soil-creep – the movement of soil down a steep slope – is common on the sides of the dry valleys, forming horizontal ridges called terracettes **below**. Where chalk downland is eroded, in quarries or by the sea, the startlingly white nature of the rock is very marked, as here **right** in Sussex, southern England.

Hand specimen The rock breaks off in irregular dusty pieces as there are no internal structures or zones of weakness. Fossils are so completely encased that they cannot be removed. Any break cuts clean across the fossil. The fossils are usually formed by silica replacement, and if there are any veins present in the rock these will contain silica too. Chalk is so fine-grained that the constituent particles are impossible to see, either with the naked eye or with a hand lens. A grayish tinge may be due to the presence of mud particles in the original sediment. A greenish tinge may be imparted by the iron silicate mineral glauconite, but only in impure chalks.

Varieties Chalk is so pure that the slightest impurity can impart a characteristic color and give it a name. White

chalk is the purest. Gray chalk has fine mud or clay particles. Red chalk contains iron minerals.

Chalk forming on the bottom of the sea may be affected by bottom currents, and may be broken up and redeposited. The broken pieces are then recemented together and may be mixed with nodules of other minerals. The resulting rock is called chalk rock and can be met as distinct beds in chalk sequences.

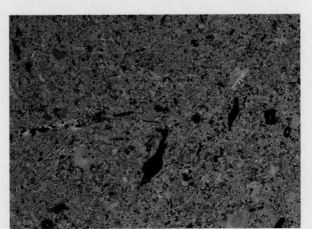

Microscopic specimen The grain of chalk is so fine that even microscopic investigation can be of little value. Indeed it was only with the invention of the electron microscope in 1932 that it was possible to identify the types of minute shells that formed the bulk of the rock.

SANDSTONE

Sandstone is a clastic sedimentary rock, with a grain size of between 0.0002 in and 0.02 in (0.02 mm and 2 mm). Anything bigger would be a conglomerate or a breccia, anything smaller would be a siltstone, shale, mudstone, or clay. Various types form in deserts, in rivers, and on sea beds.

Mineralogy

The mineralogy of sandstone depends on the rock from which the original sediment weathered, and also on the conditions of deposition of the sediment. Usually it is the most robust minerals that survive as sand grains, and that generally means quartz. Other minerals from the parent rock may also be present if the grains have not traveled far before deposition. Cementing material can be calcite, iron oxide, or more quartz.

Landscape and structures

If the sandstone is derived from desert sands, it will form a massive bed covering a very large area. Dune bedding will be present. The rock will probably have a reddish tinge due to iron oxide that formed when the original sand was exposed to the atmosphere.

River sandstone will be more finely bedded, and will probably be interbedded with other rocks such as shale or mudstone. It is unlikely to be pure and the impurities may make it shaly sandstone or limy sandstone. Current bedding may be present, as may slump bedding. It will not have the reddish hue of desert sandstone. If the cementing material is iron ore, it will be brown; if it is calcite, it will be gray. It may be leached white and full of root fragments, in which case it formed a sandbank above the water surface and had plants growing in it. Watch out for washout structures. These represent the channels of contemporary streams that cut through a sandbank and then filled up with their own sediment. They show as trough-shaped structures filled with sandstone that has a distinctively different bedding from the surrounding rock.

Sandstone formed in shallow seas may be very similar to that formed in rivers. In this case ripple marks will be more common than current bed-

Left Massive sandstone – consisting of thick beds with few bedding planes – will erode into sheer bluffs and towering crags, particularly when exposed to wind erosion, such as here in Utah.

Above Sandstone is a porous rock and holds water very well. In well-watered regions, such as France, a sandstone landscape is able to support a prolific growth of vegetation.

ding. There may be sedimentary dikes and sedimentary volcanoes – self-contradictory terms that describe what happens when a quicksand squirts up through a more rigid layer that has been deposited on top. These may be associated with slump bedding.

Sandstones formed in deep seas are quite different. They may be very coarse and jumbled, or may show graded bedding, depending on the state of the sea-bottom currents that formed them.

When studying sandstones in the field it is important to note the orientation of the strike and the angle of dip. The thickness should be mea-

sured, and measured again at a different site to see how regular the bed is. If it contains current bedding or sole marks, some attempt should be made to determine the direction of the current.

As with any bedded sequence, the folding should be noted and the orientation of the folds marked on your map. Likewise note the angle, the throw, and the orientation of any fault – faults are easier to see in bedded rocks than in unbedded. All this information will help you to understand the deformation that has taken place since the beds were laid down.

Hand specimen A smallish hand specimen should be enough to show the features you would like to study, unless the outcrop shows distinctive large-scale structures such as current bedding. If there are some really good features, like tool marks or fossil footprints, it is probably best to leave them alone so that they can be seen by others in their proper geological context. Take photographs rather than samples. The stone-rubbing technique (pp. 56–59) may be suitable for gathering a record of your discovery.

The individual grains in a sandstone are usually big enough for examination by a hand lens.

Varieties *Grit* is a coarse sandstone with very angular fragments. It was probably deposited at the mouth of a fast-flowing river.

Quartzite is a pure sandstone consisting of quartz cemented by more quartz. The desert sandstone described above is an example. The term is also applied to a thermal metamorphic rock consisting of nothing but quartz.

Greensand is a coarse quartz sandstone with feldspar and mica but containing large crystals of the green magnesium-iron mineral glauconite, formed as the sand was deposited on the bottom of the sea.

Graywacke is a coarse unsorted sandstone, usually full of angular fragments of quartz, magnesium-iron

Greywacke

Millstone grit

Penrith sandstone

inches

centimeters

minerals and others in a groundmass of much finer material such as clay and mud. This usually forms as continental shelf sediments collapse down the edge of the continental slope into deeper waters. It may show slump structures or graded bedding.

Arkose is a sandstone that has more than 25 percent feldspar and is derived from the weathering of a nearby granitic terrain.

Seat earth is a very pale leached sandstone full of root fragments. This represents an ancient sandbank on which

vegetation was growing. It is usually topped by a bed of coal.

108

Microscopic specimen

The analysis of sandstone through a microscope is a science in itself. According to the nature of the grains, you can tell if the sand from which the sandstone formed was derived from igneous, metamorphic, or sedimentary rocks, if the outcrops from which they were eroded were close or a long way off, and usually something about the conditions of deposition.

Desert sandstones usually consist of nothing but quartz. The grains will be almost spherical, having been eroded by wind action for some time. They are usually cemented together by iron oxide. Sometimes they are cemented by quartz, in which case the original quartz grains could have continued to grow after being buried. The shape of these original grains can be seen as ghostly circles in the middle of each crystal.

Grains of feldspar in a sandstone suggest that the sand was deposited in arid conditions and rugged topography, such as in a mountain desert stream. Low topography and humid conditions produce slow-moving rivers that tend to break down the feldspar before the sand is deposited.

If the quartz grains show straight extinction, the sand has probably been eroded from granite. If it shows undulate extinction – if different parts of the same grain become dark at different times – then the crystal has been stressed and is probably derived from a metamorphic rock. If the grain consists of several crystals, all with straight extinction, the sand has probably formed when a quartz vein eroded.

Likewise, quartz grains derived from a vein may have tiny bubbles embedded in them. Rock fragments are particularly useful. These are grains that consist of more than one mineral. They give a direct glimpse of what the original eroded rock was like.

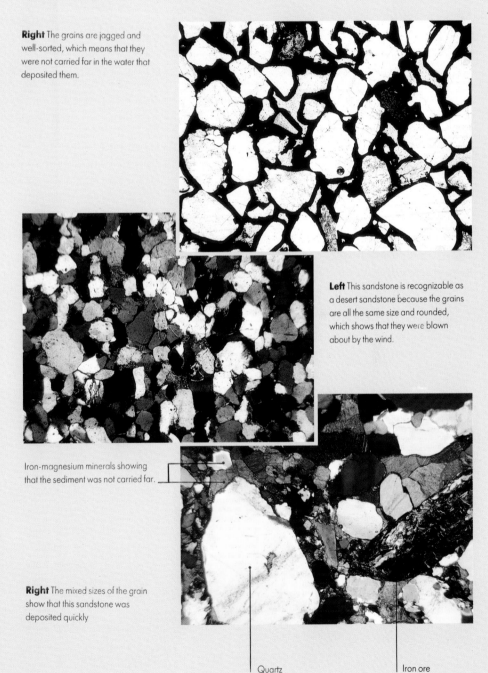

Right The grains are jagged and well-sorted, which means that they were not carried far in the water that deposited them.

Left This sandstone is recognizable as a desert sandstone because the grains are all the same size and rounded, which shows that they were blown about by the wind.

Iron-magnesium minerals showing that the sediment was not carried far.

Right The mixed sizes of the grain show that this sandstone was deposited quickly

Quartz

Iron ore

FINE-GRAINED SEDIMENTARY ROCKS

The finest sedimentary rocks are the shales, mudstones, and clays. They may be clastic or chemical sedimentary rocks, but are usually a mixture of both.

Mineralogy

The fine particles of these rocks may include fragments broken from preexisting rocks, such as quartz and, less commonly, feldspar and mica. There may be concentrations of rarer minerals such as zircon ($ZrSiO_4$).

Most particles, however, are of substances formed by the breakdown of other minerals. These include the clay minerals such as kaolinite ($Al_2(Si_2O_5)(OH)_4$) – the mineral from which china clay is made – and minerals that are akin to mica. These form a kind of a paste or an amorphous groundmass that is difficult to study.

Landscape and structures

These fine-grained rocks may form in the deep oceans or they may form in river beds, estuaries, and deltas. Each occurrence produces a different outcrop.

Those from deep oceans tend to form very thick sequences. They will probably contain fossils of deep-sea animals, or of animals that lived in the upper waters and sank after death. A soft muddy surface is very susceptible to markings, and so you may find sole marks or groove casts where blocks of debris have been dragged along the ocean floor by the currents, plowing through the top layer of ooze.

Occasionally any limy material present organizes itself into discrete layers after deposition to produce an alternating sequence of limestone and shale beds, such as the Jurassic Lias of England.

Shallow-water fine-grained sediments are usually found interbedded with sandstone and other sedimentary rocks, often in a cyclical sequence (see pp. 58–61). When this happens, the fine-grained rock, being much softer than the sandstone, will be eroded first, leaving the harder beds jutting out as ledges. Fossils are often abundant, and the friable nature of the rock makes them easy to collect.

Nodules may be present. These form when a chemical in the sediment, such as silica, carbonate, or pyrite, gathers in a discrete mass and forms a lump at the same time that the sediment is turned to rock. They usually form flattened spheroids that lie in the plane of the bedding. Septarian nodules are worth looking for. These have a radial pattern of cracks filled with yet another mineral.

As with any bedded sedimentary rock, measure the thicknesses of the beds, determine the direction of strike and the angle of dip. Note any folds or faults.

Right The fine-grained sedimentary rocks are usually very soft compared with any other rocks in the area. Where clays, mudstones, or shales are interbedded with limestone, the beds of limestone are left protruding as shelves or ridges while the intervening fine material is washed away. This gives a deceptive step-like appearance to any cliff, although its unstable nature forbids any climbing.

Hand specimen Fine-grained sedimentary rocks are so soft that specimens are easily obtained.

When the rock can be split easily into thin brittle sheets, it is a shale. When it breaks into lenticular flakes, it is a mudstone. When it has little internal structure but is plastic and slippery when wet, it is a clay. If the distinction is not obvious when you hammer off a hand specimen, then apply the blade of a penknife and see how the specimen splits.

Fossils are likely to be present and these are most easily seen in shale because it spilts along the bedding plane where fossils are usually aligned. Fossils of marine animals are found in deep-sea shales; freshwater shellfish and plants lie in shallow-water shales.

Very dark marine shales can be rich in carbon, showing that they were deposited in a region deficient in oxygen (otherwise the carbon would have been incorporated into limestone or given off as carbon dioxide). Fossils in such shales may be of animals that suffocated after being swept into these regions by currents. The lack of oxygen leads to the formation of iron pyrites, and this may be present as crystals or as a replacement mineral in fossils.

Clay

Mudstone

centimeters

inches

Shale

Varieties Within the divisions of shale, mudstone, and clay, there are many different types. They are usually given names that are based on their economic importance.

China clay or kaolin is a white clay formed by the decomposition of the feldspars in granite.

Fuller's earth is a clay made from very fine volcanic ash, used for removing grease from wool.

Alum shale contains minerals that can be worked for alum.

Microscopic specimen Don't bother. The constituent fragments are too fine for the technique to be of any use.

SLATE

Slate is a low-grade metamorphic rock formed by the regional metamorphism of a fine-grained rock such as shale or volcanic ash.

Mineralogy

As with most metamorphic rocks, the chemical composition of slate is similar to that of the original rock, but the constituents have been rearranged into different minerals. The most significant minerals produced are those that form thin flat crystals, like mica or chlorite, developed from the original clay minerals of the rock. Quartz is likely to be the other major constituent.

Landscape and structures

Being regional metamorphic rocks, slates tend to be found in mountainous areas. The rock cleaves into thin sheets because of the alignment of the minerals, and so it tends to weather in a splintery manner. Erosional forces open cracks in the rock parallel to the cleavage plane.

When the cleavage is perfectly flat, the slate can be quarried and used for roofing material or for billiard tables. As slate is found in large masses, slate quarries tend to be huge, eating away entire mountains in such rugged areas as North Wales and the Appalachians of eastern North America.

The direction in which slate splits has nothing to do with the bedding of the original rocks. In fact, in some slates you can see the original bedding cutting across the cleavage plane. The cleavage is entirely due either to the alignment of the minerals – true cleavage – or the alignment of closely spaced microscopic folds – false cleavage, fracture cleavage, or strain-slip cleavage. The alignment corresponds to the direction of applied pressure as the rock metamorphosed.

Through strain markers it is sometimes possible to see the extent of the deformation. These are recognizable objects that have been distorted by the forces that changed the rock, and can include fossils and spherical pebbles.

Since you will usually be studying slates in mountainous areas, you should be sure to wear suitable clothes and shoes.

Note the direction of cleavage, and whether or not it is even for as much of the outcrop as you can see. If the direction varies from place to place, there may have been more than one phase of deformation – mountain-building movements metamorphosed the rock, and then more mountain-building movements distorted the metamorphosed rock.

Right Slate tends to form mountains, being produced by the regional metamorphic processes that accompany mountain-building. The rock was once valuable as a roofing material and for other industrial purposes, and so slate areas tend to have quarries and quarry towns in them, such as here in north Wales. Changing economic factors mean that now many of these quarries are disused and the towns run down.

Hand specimen Slate tends to be a fairly even rock, and being fine-grained a small sample is usually representative of the whole emplacement. Its color is usually a dark gray but, according to the precise mineral content, it can be green, blue, or reddish-brown. A hand lens may reveal the mica flakes that give the rock its cleavage, but often they are too small.

Varieties As with all rocks that have a commercial significance, there are several types recognized by their appearance. They vary in color according to their mineralogical makeup and some may have largish crystals scattered throughout, giving the slate a spotted appearance.

Similar rocks If slate continues to be subjected to the forces that produced the metamorphic effect, the grade of metamorphism will increase and produce phyllite – the next stage in the process. Phyllite is very similar to slate but the crystals are much larger, the shiny plates of mica being visible to the naked eye.

Even further pressure will develop the next stage of rock – schist.

inches

centimeters

Welsh slate

Microscopic specimen Through the microscope it is easy to see whether the cleavage is true or false. The former will show itself by the alignment of minerals, while the latter will present a wavy texture in the fine groundmass. Strain markers are visible on the microscopic scale as well. Tiny robust fragments and crystals that have resisted the metamorphism may be at the centers of eye-like structures. The matrix appears to have been pulled out away from the fragment in the direction of cleavage and the spaces so formed filled in by chlorite or quartz, like a miniature augen (p.54).

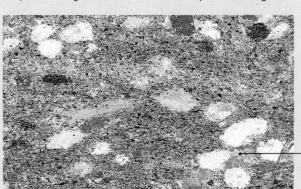

Chiastolite – an unusual form of andalusite (aluminum silicate) crystals in cross-shaped arrangements

SCHIST

Schist is a medium-grade regional metamorphic rock. It is found in ancient rock systems, where there were once mountains which have subsequently been eroded to stumps or to flat plains.

Mineralogy

The minerals in schist vary greatly, depending on the nature of the original rock and the circumstances of the metamorphism. Mica is nearly always present. A large proportion of mica suggests that the original rock was a fine-grained shale or mudstone. Less mica and more quartz and feldspar suggests a parent rock of sandstone. The presence of calcite shows that the original rock was limestone. Talc in a schist reveals that it formed from a basic igneous rock.

New minerals appear under conditions of increasing temperature and pressure. Geologists can identify these minerals and use them to determine the grade of metamorphism. Mica develops with light metamorphism. Then comes garnet mixed in with the mica. Then staurolite ($FeAl_4Si_2O_{10}(OH)_2$) – a yellow-colored mineral often with a cross-shaped twinned crystal. Then kyanite (Al_2SiO_5), which may form pale blue crystals. The final grade is signaled by the appearance of colorless fibrous crystals of sillimanite (with the same chemical formula as kyanite). Any further stress and the rock turns to gneiss or begins to melt.

Landscape and structures

A schist terrain has been greatly eroded. When you see outcrops of schist, you can tell that vast volumes of rock have been worn away in order to reveal them.

Like slate, schist is made up of flat minerals that have been aligned parallel to one another, and, like slate, it splits and weathers along these planes. However, schist cleavage planes are very irregular, undulating, and twisted. The whole range of tectonic structures – anticlines, synclines, thrusts, faults – can be seen affecting the cleavage. Mullions (pp. 54–55) may develop, in which case mica may be formed around the outside of each of the cylindrical structures. The contorted nature of any outcrop will show the complex history of the area, with one phase of metamorphism and distortion superimposed on another. The most obvious direction of deformation is the one that was produced last. It may obliterate all the early ones.

Left Jagged mountain peaks are typical of schist terrain. The cleavage planes caused by the layers of mica in the rock mean that it is vulnerable to physical weathering and splits easily. Schist mountains are easily attacked by frost and form spectacular rugged scenery. The micas themselves may cause schist rocks to gleam and sparkle in the sunlight.

Hand specimen Hammer a specimen from the outcrop. It will split away along the cleavage plane. The presence of this splitting direction is so typical of schist that the term schistosity may be used about any rock, schist or otherwise, that shows it. In the United States the term foliation is used instead, but in other parts of the world foliation means the separation of minerals into discrete bands, as in gneiss.

The broken face of a hand specimen of schist will sparkle in the sun because of the mica crystals. The crystals may be big enough to be seen with the naked eye, but if not you will definitely see them with a hand lens. The other minerals, particularly the index minerals that show the grade of metamorphism, should be looked for too. Garnets will be readily recognizable by their deep red color and their other properties (p. 29), and kyanite should be obvious

by its light blue color. The other index minerals may be difficult to detect without trained help.

Varieties The many varieties of schist are the result of differences in the parent rock, and the grade of metamorphism.

Mica schists are the most common.

Calc-schist, with a high proportion of calcium, is derived from the metamorphism of limestone.

Greenschist, with a high proportion of the green mineral chlorite, is derived from basic igneous rocks.

Quartzo-feldspathic schist comes from metamorphism of sandstone.

Microscopic specimen A microscopic view of schist should show the schistosity by the alignment of the minerals, particularly of the micas. The presence of garnet should be obvious, and sometimes the well-formed garnet crystals are encased in an eye-shaped lens of quartz. Staurolite, if present, may show itself by large cross-shaped crystals.

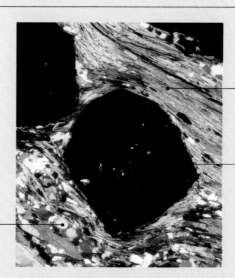

Mica – brightly colored and showing the strain of the formation

Garnet – isotropic therefore dark under crossed polars

Small quartz grains

GNEISS

Gneiss (pronounced "nice") is a coarse-grained regional metamorphic rock. Heat as well as pressure is involved in its formation and this is the most metamorphic rock known. Often it is impossible to tell what the original rock was.

Mineralogy

As a rule the mineral content of gneiss is similar to that of a very high grade schist. The minerals, however, are much coarser and separated into distinct bands.

Landscape and structures

As with most regional metamorphic rocks, outcrops of gneiss may stretch over many hundreds of square miles or kilometers. They are only formed in the depths of continents in areas of intense mountain-building activity, so its presence at the surface indicates that a great deal of material has been eroded away. An extensive area of gneiss is often referred to as basement because it is inferred that this is the kind of rock that underlies all other rocks of the continents, forming the foundations of the continents themselves. Some of the oldest known rocks are gneisses.

The most obvious structural feature is the banding of the minerals. Each mineral has separated out into a discrete layer – foliation – and these layers have been greatly twisted and contorted by the pressures that formed the rock. Sometimes a layer of mineral will have been pulled out into a series of lenses. Part of the rock may have melted into granite, and the molten material may have been squeezed into cracks throughout the outcrop. This would give veins of quartz and pegmatite, or a complex mess of metamorphic and igneous rock – migmatite ("mixed rock"). A quartz vein so formed is very sinuous, like an aerial photograph of a meandering river. Unlike in bedded sedimentary rocks the layers do not usually erode at different rates – an exposed face of a gneiss outcrop may have little relief.

Above We see gneiss in places where ancient mountain ranges have been worn down to their bases.

A landscape of gneiss, such as here in the Scottish isles, tends to be rounded and smooth. This is not just because erosion has been acting on it for so long, but also because the rock tends to be of a fairly even hardness with few of the cleavage planes we associate with other forms of regional metamorphic rock.

Hand specimen When you hammer off a specimen, it will not usually split along the foliation plane, such as schist does. The resulting specimen will be irregular and cut across the different bands.

The minerals in gneiss form crystals that are big enough to be seen with the naked eye. Light-colored layers of quartz and feldspar crystals alternate with darker mica-rich, amphibole-rich, and pyroxene-rich layers. The crystals can be so big that you can identify them from their physical properties (pp. 26–27). Sometimes the mineral bands and structures are so big that it is easier to study them in outcrop than in hand specimens.

Varieties Gneiss is usually classified according to its most abundant mineral.

Muscovite gneiss is named after the white mica muscovite. It is one of the

inches

centimeters

most widespread of gneisses and is probably derived from sediments.

Hornblende gneiss is named after its amphibole mineral and may be derived from basic igneous rocks.

Injection gneiss is another name for migmatite.

Granite gneiss is a metamorphosed granite, or the transitional stage between metamorphic and igneous rock. It is difficult to tell where extreme metamorphism stops and partial melting begins.

Similar rocks As noted above, gneiss grades into granite, and it can be difficult to tell one from the other. As a rule, however, gneiss shows foliation while granite does not.

Microscopic specimen A big crystal found in gneiss will show evidence of having been stressed – indicated either by an internal structure that appears to have been pulled and twisted about, or by such optical properties as the extinction angle differing across its area. Large crystals of feldspar may have broken up into a mosaic of finer grains.

Band of finer crystals – contrast with quartz produces banding in rock

Interlocking grains of quartz

VILLAGE GEOLOGY

Geology is everywhere. Walk through the countryside and everything that you look at is influenced by the geology, from the kinds of plants growing by the roadside and in woods, to the crops planted in the fields, to the rocks that go into the local buildings. A brief examination of human activity in a particular area can tell you much about the geology.

Quarries and earthworks

Where the local stone is quarried, the quarries will usually be quite visible. Ask permission, though, before you look at a quarry, or better still, obtain written approval from the local office. The address and telephone number are usually displayed somewhere near the entrance. The roads leading to the quarry can yield valuable information. Pieces of the stone invariably fall from the backs of trucks soon after they have been loaded, and good samples may be obtained from the roadside. Spoil heaps from old abandoned excavations are often valuable sites for collecting.

Other excavation work includes cuttings for highways and railroads. Since trains cannot usually operate on steep slopes, the bed of the track must be as flat as possible and this means building embankments across valleys and cutting deep gorges through hills. With the decline of parts of the railroad systems in many countries there is a wealth of abandoned railroad cuttings, most of which show good sections through the local geology. As with any steep section you must beware of falling rocks. Roads do not need to be so flat and so the cuttings tend not to be so deep. Often they are planted with grass and trees soon after they are made and so the geology is soon obscured. If you want to look at geology in a road cutting it is best to look at a new one. But be cautious of traffic!

Agricultural geology

Freshly plowed fields may uncover the occasional fossil from underlying bedrock. Walls around the fields may be made from the local rock. These,

Right The pyramids of Egypt were built on the banks of the Nile. The main materials used were poor-quality, foraminifera-rich limestones quarried locally. Each pyramid had an outer coating of blocks of much finer limestone quarried farther up the Nile. The chambers and passageways were lined with blocks of granite quarried at Aswan 500 miles (800 km) upstream. These different rocks were extracted from riverside quarries and brought down the Nile to the site by rafts.

Left An old settlement situated in an area where the local geology consists of a good building material, such as this village in the limestone-rich Dordogne valley in central France, is likely to be built from stone quarried locally. Some of the more minor components, such as the roofing tiles, may have come from some distance away.

however, will probably be covered with moss and lichens, and it would be neither polite nor legal to hammer off samples.

Buildings

Old towns and villages tend to be built of local stone. Originally this was just because the local stone was handy. Nowadays, in many countries, regulations may insist that local stone is used in new buildings so as to maintain the character of the area. Sometimes this shows itself as the deliberate inclusion of spectacular local fossils, such as big ammonites or dinosaur footprints, in decorative parts of the new buildings. The geology of masonry can be very useful to the archaeologist. The types of stone used in old buildings can tell much about an area's history.

Archaeological and geological cooperation

The village church in Brixworth, Northamptonshire, in the English Midlands, is built of many different stones. Geologists, asked by local historians to analyze the masonry, deduced that much of the original 7th- and 8th-century brickwork and sandstone blocks came from demolished Roman buildings in Leicester in the next county to the north and Towcester in the south, indicating possible trade routes in medieval times. Locally

quarried stones were not used, suggesting something about the medieval economics of reusing old building stone from far away rather than fresh, locally available materials. Later additions were built of local sandstones and tufa in the 11th, 12th, and 14th centuries, indicating that it had then become more economical to work with local resources.

URBAN GEOLOGY

The location of all human habitation is a result, directly or indirectly, of geography, and this is in turn the result of geology.

A city may arise as a center of communications, where an east-west route along a narrow coastal plain meets a north-south route following a river valley through mountain passes. Another may develop because an inlet of the sea gives a good harbor, from which the product of a nearby mineral mine can be exported. These are all geographical, and therefore geological considerations.

Like a village, a city would first have been built of the local stone. The oldest buildings in the city may still show this. However, with the development of communications and transportation systems, local materials are becoming less and less relevant in determining urban architecture.

Take a walk down a city street. Most of what you see will be concrete and steel – derived from limestone and iron ore – and a long way from the minerals that supplied them. However, much of the decorative material that you see will have a geological interest. Floors of banks, or the walls of insurance buildings – institutions that try to put on a friendly public face – may have veneers of attractive stone such as marble, polished limestone (also often called marble by builders), mica-rich gneiss, porphyritic granite, and so on.

Where older buildings still stand, the polluted city air may have severely eroded the stonework. Such buildings are often cleaned to preserve the city landscape. When this has happened to a limestone building, have a close look at the surface. An outer coating may have been eroded by atmospheric acid, leaving fossils etched in relief.

Geology and history

Millstone trade routes

In their day the Romans ruled the countries of the Mediterranean. We can trace their trade routes by looking at the occurrences of millstones. Vesicular lava was the preferred material, because the bubbles in it ensured that there would always be a rough surface when the millstone became worn down. The best millstones came from outcrops at Mulargia in Sardinia and Orvieto in Italy. They have been found in archeological sites in Spain and North Africa. It looks as if Romans transported them to the wheat-growing areas of Spain and North Africa (climates were different in those days) in the ships that returned laden with wheat. Other lavas, from Sicily, Morocco and the Aegean islands, were used more locally. These millstones do not seem to have been so valuable as they were not transported so far.

Continental drift reversed

Most of the old "brownstone" buildings of New York are made of Triassic sandstone quarried in the nearby Connecticut River Valley. Some, however, are built of Scottish sandstones brought across the Atlantic from Dumfriesshire in the 19th century. Sailing ships returning empty to the United States had to take local rock as ballast and dumped it when they reached harbor. Now it is very difficult to tell buildings made of the one from buildings made of the other. Both sandstones were formed at the same period under the same circumstances in the same general area. In Triassic times a mountain area crossed the northern supercontinent. Mountain streams brought debris down from these mountains into a desert region to the south, where it was deposited as shallow water and desert sandstones. In the 200 million years since their deposition, plate tectonics have opened up the North Atlantic Ocean between their two sites and they are now 4350 miles (7000 km) apart. Civilization has now, inadvertently, brought them back together in one city.

MAPPING

The science of geological mapping was established by Charles Lapworth, the 19th-century British geologist.

About fifty years earlier, in 1815, the English canal engineer William Smith had drawn the first geological maps, in which rocks of different ages were depicted in different colors. He used the different fossils found in each rock to identify it, and was the first to do so. This technique is now called biostratigraphy and is the mainstay of modern geologic dating. Lapworth later applied the scientific study of rock structures to geological mapping and the science become more or less as we know it today.

It is quite feasible to draw your own map (see following pages), but there are other types of maps used by geologists, and these are usually of a somewhat specialized and technical nature.

The tectonic map

Often the structure of an area is more important

Left William Smith published the first small-scale geological maps in 1815 and established the tradition of identifying different rock types by different colors. Modern geological maps – both large-scale and small-scale – are much more sophisticated than Smith's but adhere to the same colorful tradition.

than the rocks themselves. The structure may be so complex that if the information were included on a conventional geological map the crowding of the information would just be too confusing. As a result, a tectonic map may be drawn. On this only the structural trends are shown, such as the fault lines and the axes of the synclines and anticlines. If the rocks are identified, then they are only shown by coarse classifications – Mesozoic as opposed to Triassic, Jurassic, or Cretaceous.

The facies map

The term facies is one often used by geologists, yet even they find it difficult to define. Facies means the whole character of the rock – its type, its thickness, its texture, its fossils, its mode of formation, its feel, its taste, its smell – in fact the gestalt of the rock.

A facies map takes a particular bed, or a particular horizon that represents a single period of time over a large area, and plots the different rock types – the different facies – across that area. It is a study widely used in the search for oil, and so it has become very complex and technical.

Usually the information for compiling a facies map comes from boreholes, as it is unlikely that the one horizon will be exposed over its whole extent. The more boreholes that are drilled, the more accurate the map will be.

The paleogeographic map

A facies map can be used to construct a paleogeographic map. This is a map of land and sea areas, mountain ranges, deep or shallow water, and all kinds of other geographic features for a particular period of geological time. These can be made very realistic and exciting – showing ancient swamplands where dinosaurs wallowed, vanished reefs where trilobites swarmed, lost islands where pterosaurs nested – but those with a great deal of detail tend to lack accuracy. Paleogeography is an inexact science, and there is a great deal of speculation in placing extinct landforms on a map. Usually we can say no more than, "There was a shallow sea here, giving a chemical deposition of limestone, while over there was a river delta producing sands with current bedding and layers of coal." Even then we cannot be sure if the two existed at precisely the same period of geological time.

MAPPING 1

Every geological map starts as a published topographic map, on which the landscape features are marked and the heights given as contour lines. Such a map is fine for finding your way about, but geologists want to know what the surface rocks are made of and what lies beneath them.

If the geology of an area consists of undeformed sedimentary rocks, lying horizontally, mapping is no problem. The outcrops of the various beds are parallel to the contours. This, however, is a very unusual occurrence. Even if the beds are not deformed in any way they are usually dipping. As a result, they will outcrop at different heights at different parts of the landscape.

The important concept here is the strike (see pp. 46–47), the line that a bed makes with a horizontal plane. The angle at which the bed dips can be found by drawing in the strike lines at various heights and using the information to draw a cross section.

Map A This map shows a region of hilly terrain in which a bed of coal outcrops. Wherever it outcrops it does so between the 500 ft and 600 ft contours. From this we can deduce that the bed is horizontal.

We can prove this by drawing a cross-section. First, select a suitable line along which to draw the section, in this case line A-B. Now place a straight edge of paper along the line and mark off the points where it crosses a contour. Prepare a grid on which to draw your section – a grid on which the various altitudes are marked. Now transfer the contour points from your edge of paper onto the grid, and draw in the topography like a graph. Do the same for the outcrop of coal. You will see that the bed of coal is horizontal.

Map B Now we have a map of hilly terrain in which the bed of coal does not outcrop at the same elevation everywhere.

First we see where the outcrop crosses a particular contour, in this case the 600 ft contour, and mark it with a cross. Then we see where else it crosses the same contour and mark these other points with crosses. A line that joins up all these crosses is called the strike line.

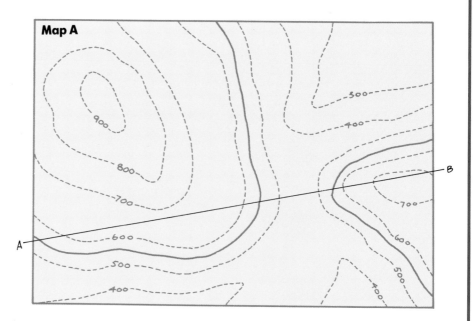

Map A

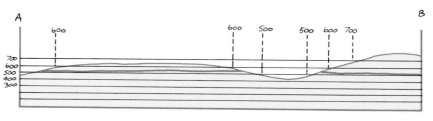

Having drawn this strike line we now draw another strike line for the 500 ft level, and then for the 400 ft level.

When we draw our cross section we usually draw it at right angles to the strike lines. Draw in the topography as described in map A, and then, using the same paper edge technique, mark off the positions of the strike lines and connect them. From this you will see the dip of the bed directly.

Note: Do not be tempted to measure the angle of dip from this cross section, unless you have first ensured that the vertical scale is the same as the horizontal.

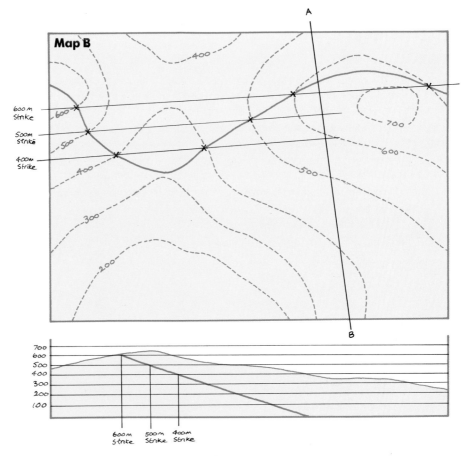

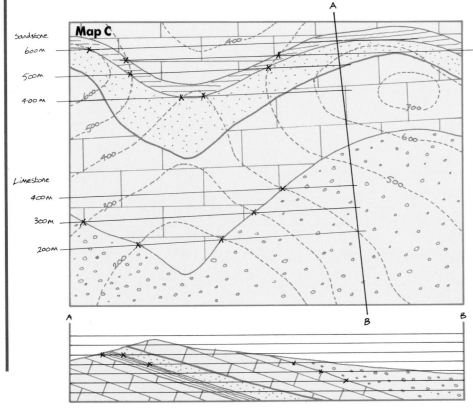

Map C This is the same area as that in map B, but the beds above and below the coal are also shown. Using strike lines for each of these, we can draw a cross section that shows the entire geology.

125

MAPPING 2

That was easy. The beds had a constant dip and were undisturbed. Often, however, there are folds, faults, igneous intrusions, unconformities, and all manner of other complexities to bedevil the geologic mapper. Let us take them in a logical sequence.

This may seem a bit laborious but after a few tries you will get your eye adjusted – you will be able to judge a scene in three dimensions and dispel any ambiguity in drawing the strike lines.

We find that we have drawn three sets of strike lines: one descending to the east, the next rising, and the third descending once more. The bed is therefore folded into a syncline and an anticline.

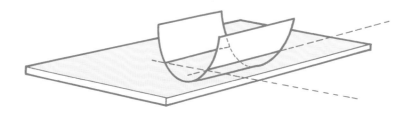

Map D A deep gorge has limestone outcropping high on its banks at each side. We are trying to map the base of the limestone in order to produce a cross section. The outcrops seem to be irregular. There is, for example, more than one strike line that can be drawn for the 500 ft level, since the base of the bed cuts the 500 ft contour several times in the southwestern corner, at points w, x, y, and z. We could try to draw the strike lines w-x and y-z, but we could see that this is incorrect because the base of the bed obviously sags between w and x (and the strike line is supposed to mark the line along which the bed is horizontal). Another guide is that the lines w-x and y-z are not parallel to each other, whereas w-y and x-z are, suggesting that these represent the true strike.

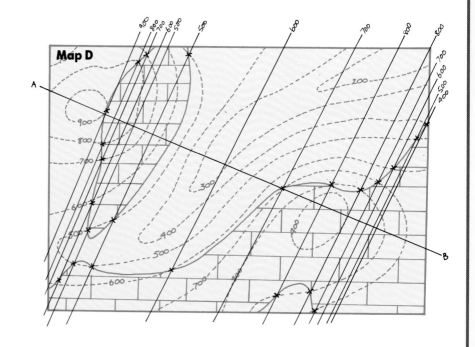

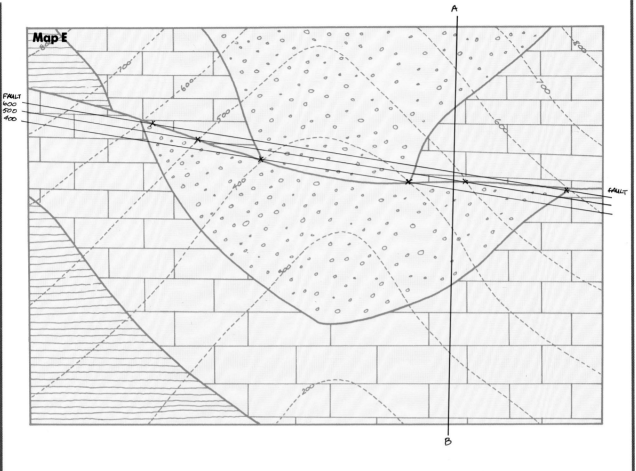

Map E

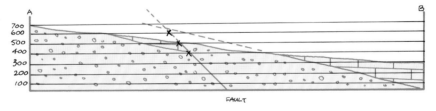

Map E A slope in a valley has an outcrop of coal between beds of sandstone and limestone. A fault runs right across the area, displacing the geology.

The first thing to do here is to analyze the fault. Draw strike lines for the plane of the fault and put it in the cross section. This will give the angle of the fault plane.

This can be expressed as the dip of the fault plane measured from the horizontal – like describing the dip of sedimentary beds, or else as the "hade" which is measured from the vertical.

Once the fault is drawn then the rest of the cross section can be prepared. Regard it as two separate

maps, one for each side of the fault. By projecting the bed of interest, ie the coal seam, to meet the projected line of the fault, the vertical throw of the fault can be estimated. In this instance there is a reverse fault with a throw of 200 ft.

127

MAPPING 3

Unconformities occur when a sequence of rocks is lifted above sea level and then eroded. An incursion of the sea then deposits a fresh sequence of beds on top of the remains of the first, and the older beds usually meet these new ones at an angle. An unconformity may be quite evident from a geological map by the way that certain of the mapped beds seem to disappear beneath others.

Map F Outcrops of sandstone, shale, mudstone, and limestone outcrop in a NE–SW pattern, yet in the higher lands they are all overlain by a bed of conglomerate and sandstone. Mapping and the drawing of a section reveals the presence of an unconformity. As with the presence of a fault, deal with the unconformity first and draw it on the section. The rest may be treated as two maps. For the sake of clarity, only the strike lines for the unconformity are shown here.

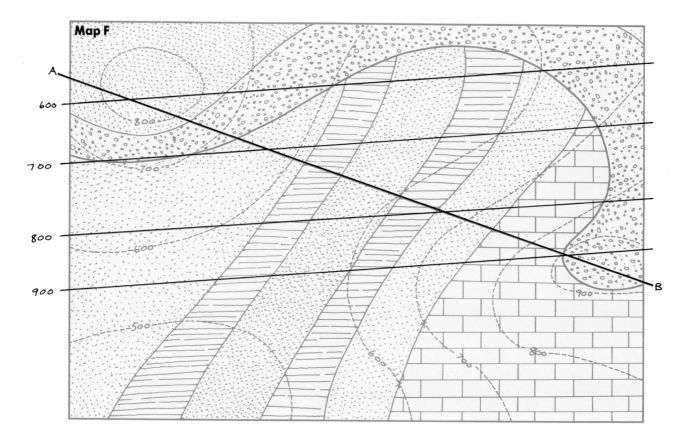

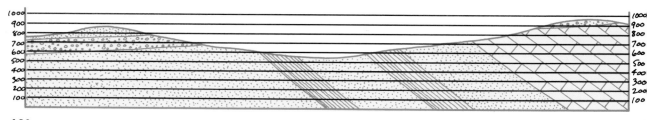

Map G Igneous rocks can cut across the grain of the land, complicating any map work.

The map shows an area that has a sequence of sedimentary rocks – limestone, shale, sandstone, and mudstone – but there are also outcrops of igneous rocks as well – masses of dolerite. Some seem quite conformable with the grain of the land but others are in structures that cut right across. When the section is drawn we find that there are two different types of intrusion here. One behaves just like a sedimentary rock, lying parallel to the sedimentary rocks of the sequence. It is, therefore, a sill. The others cut right across the geology and the topography in straight lines. No strike lines can be drawn, and so beneath the surface they must be vertical. This makes them dikes. The dikes are confined to the beds below the sill – there are none in the mudstone that overlies it. It can therefore be inferred that these dikes are feeder dikes, originally conveying the molten material that then spread along the bedding plane to form the sill.

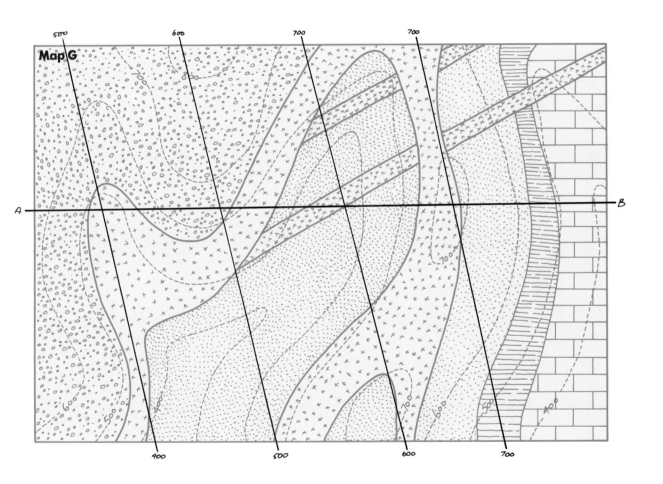

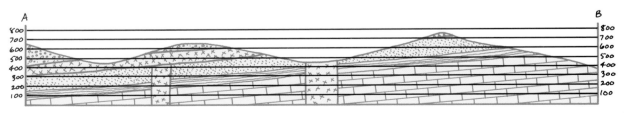

129

MAPPING 4

The maps at which we have been looking must show a terrain of bare rocks. In each case the whole of the rock sequence is visible at the surface: a geologist's dream.

That real life should be like that! In practice nothing is as simple. The rocks weather and break down. Rubble fills the hollows. Broken rocky material is intermixed with decaying vegetable matter and produces a covering of soil – very useful for the farmer, but a nuisance for the geologist. Highways and houses are built over the land. In the end all we are left with are odd exposures in disused quarries, stream beds, and railroad cuttings, and from these we must map the whole area.

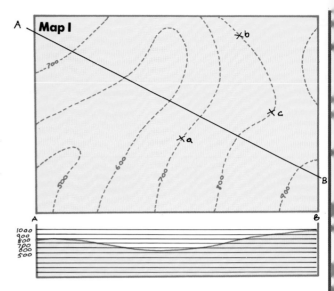

Map H A coal bed (we are using coal beds in these examples because they are conveniently thin, and have a certain economic attraction) outcrops on a hillside crag at *a*, in the gorge of a stream at *b* and also in a disused quarry at *c*. We have enough information here to draw two strike lines, one at 500 ft and one at 600 ft. Using these strike lines we can continue to trace the outcrop of the bed across the landscape in areas where it is hidden beneath topsoil.

Not far below it in the rock sequence lies a thin bed of iron-rich sandstone. It only outcrops in the stream bed at *d*. Plot its outcrop across the rest of the map.

Now, using the techniques that have become familiar, draw a cross section of the area.

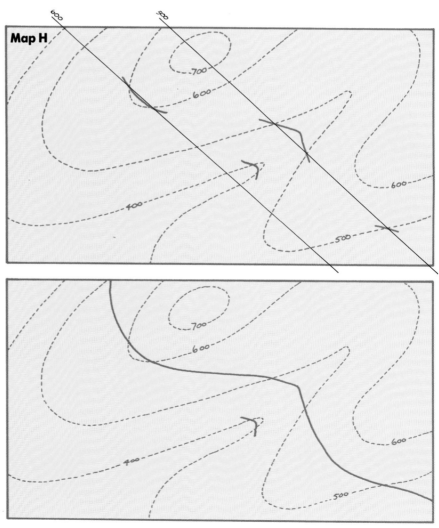

Map I Sometimes geologists do not even have outcrops to guide them. All they have are the results of borehole drillings.

During a drilling program the engineers met a bed of coal at various depths at various sites. In borehole *a* they found it at a depth of 100 ft. In borehole *b* it lay at a depth of 200 ft. In borehole *c* they found it at a depth of 300 ft. (You were able to tell the engineers that they had found the same bed of coal, as the conforming bed of shale that overlay it had the same shellfish fossils at each place.)

You are asked to see if the bed is likely to outcrop anywhere in the area of the map, and if it does, draw it in and make a cross section.

The difference here is that the known occurrences of the coal are underground, but you can still attempt to draw strike lines. By subtracting the depth at which the coal was found from the elevation of the drill head you can find the height of the bed at each place. Thereafter the procedure is the same as that for a series of limited exposures as on map H.

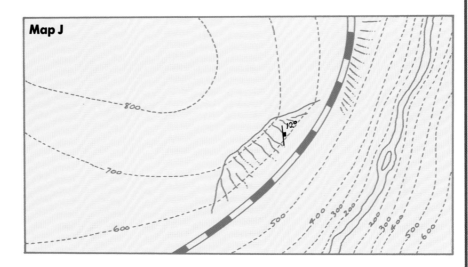

Map J

Map J It may be that your limited exposure does not have enough information to allow you to draw a strike line directly. In that case use the clinometer that you made on p. 47, and measure the angle of dip. Then you can draw the bed dipping at that angle on the cross section (ensuring that the vertical scale and the horizontal scale of your cross section are the same). From the cross section you can determine the strike lines and apply them to the map to plot the outcrop. On the example, the only exposure of the bed in which we are interested lies in a railroad embankment near the top of a gorge. Its strike runs west of north and it dips 12° to the east. We have taken the calculation to the stage in which the strike lines are drawn. Now fill in the outcrop in the gorge.

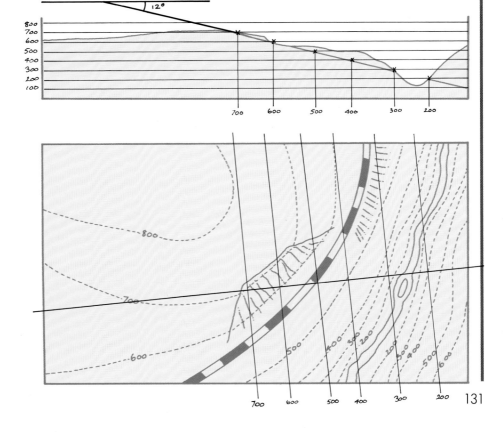

MAPPING 5

Now you have enough knowledge to go into the field and accurately map. Consider the landscape shown. In an area such as this you can tell the best places to look for rock exposures – the crags, the stream beds, the old quarries. Your first action would be to sketch it, drawing in as many of the geological details as possible.

The field sketch Such a field sketch would show the relationships of the various features to one another, and how the topography integrates with the geology. Immediately you may be able to get some idea of the geology, for example if the hills consist of scarp and dip slopes the general dip of the rocks may be obvious. The sketch is an aid to your memory, and an adjunct to your field notes. It may be done in your notebook or on a sheet of paper clipped to your map holder. Cover it with a transparent sheet held in place with an elastic band so that it can be lifted off easily when you want to make notes.

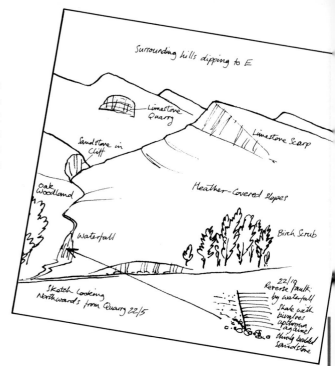

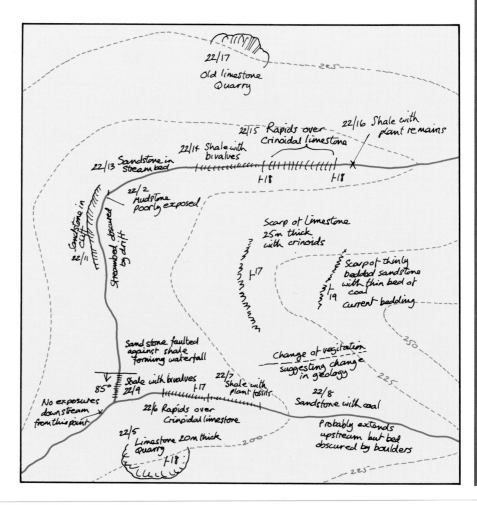

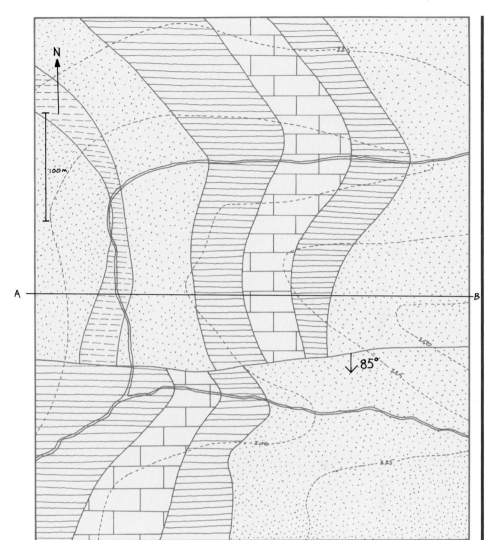

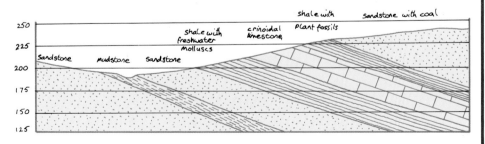

The field map Your field map will start life as a large-scale (eg 1:10,000) topographic map of the area, purchased at a specialist map store or obtained directly from the map publishers. You should take more than one, since sheets of paper tend to be somewhat vulnerable in the face of field conditions. Keep one at home on which to draw up your final map. Write on as much information as you can in the field. Number each outcrop and refer to it by that number in the detailed description in your field notebook. If you take specimens they should be given this number also.

From the information given on the examples of a field sketch and a field map, draw the strike lines and produce a geological map and a cross section of the illustrated area. If you can manage this, Lapworth would have been proud of you.

133

Labels visible on specimens:

NAME/DESCRIPTION
GNEISS
LOCALITY HEBRIDES
DATE 4/9/79   G.R. NZ 010298   REF. NO. 14/37

NAME/DESCRIPTION
HALITE (ROCK SALT)
LOCALITY CHESHIRE   G.R. ST 910423
DATE 4/11/0_   REF. NO. 8/5

NAME/DESCRIPTION
MUSCOVITE M
LOCALITY SHETLAND
DATE 4/3/87   REF.

VEIN
HORIZON
LOCALITY WALES
DATE 2/3/91

Above and right Your specimens should be kept in sturdy drawers. Each should be separated from its companions, and this is easily done by keeping them in individual cardboard trays. If the drawers are not dustproof, it is best to keep the specimens in plastic bags. Each tray or bag should contain a card with all the relevant information about the specimen, and this should be given a catalog number also inked onto the specimen itself in case the two become separated.

PREPARING AND CURATING SPECIMENS

As a practical geologist you will inevitably build up a collection. Before long you will have mineral crystals that are particularly attractive, rock samples that illustrate some geological principle more beautifully than any you have ever seen in a book or in a museum collection, fossils that bring back the memory of the exciting scramble over wave-slicked rocks beneath a cliff to get them before the tide came in, and which bring vividly to life the strange creatures of the past. At the beginning these will lie on coffee tables and windowsills, impressing visitors but gathering dust and being knocked around during hasty housework. Soon you will have so many that you will want to put them together in an organized collection.

Preparation

Home from the field, and you will have your specimens and samples securely wrapped in newspaper and marked with a number that will be keyed into the appropriate part of your field notebook (pp. 48–49) and your map (pp. 124–133). The first thing you must do is sort them all out as soon as possible while the memory of your expedition is still fresh.

Cataloging

Once you have identified your specimen – mineral, rock, or fossil – you must catalog it. Mark it with a spot of white enamel paint and then write on a number with permanent ink. This number will refer to an entry in your rock register – a notebook in which you record your collection – or to a file card. The register or card index will contain details of what it is, where it was found, the date on which it was found, what the surrounding rocks were like and anything else that you think is important. This information could equally be stored on a simple computer database.

Storage

Geological specimens are usually quite robust and it is tempting to let them lie naked on shelves or wrapped in the newspaper they came in. This is not a good idea since they will gather dust, and dust always spoils the appearance. Close-fitting drawers make good storage places for geological

135

Right Visiting a geological museum can be quite daunting. How could the amateur aspire to collect such wonderful specimens and to display them with such panache? However, your collection will improve over time, and you can be inspired and learn from the museum's expertise in presenting the specimens to their best advantage. The time will come when you will have at least one better example of a specimen than you have ever seen in any museum – and you will display it with pride.

Where to work

A workshop or laboratory is useful. The minimum requirements for this are a strong table, a sink with running water and a drainboard, an adjustable lamp, a selection of tools such as dental instruments and fine chisels, sieves for analyzing soft material, and plenty of storage space. Bookshelves containing your reference books should be to hand.

specimens. Ideally you should have a custom-made cabinet built by a carpenter who knows your needs, or purchase steel geological cabinets from a scientific supply house. Wide shallow drawers are best. They can be about 4 in (10 cm) deep – that should be deep enough to accommodate the regular size of hand specimen. Each specimen should be placed in an individual cardboard tray – easily made or purchased – which will keep each one separate. Delicate specimens can be laid on a bed of absorbent cotton or tissue paper. A label should be placed in each tray, summarizing the entry in the rock register or file index.

Display

Your prize specimens – your slab of schist, its cleavage plane sparkling with the mica minerals

removed by flaking away the surrounding shale with dental instruments. Sometimes a large fossil can be removed from its matrix by placing it in a domestic oven for a while. The fossil and the rock, if they are made of different materials, may expand at different rates and the one may come away from the other. Some fossils are made of iron pyrites, and this tarnishes quickly upon exposure to air. This process afflicts the cubic crystals of iron pyrities as well. The only way to stop this is to expose the specimen to ammonia vapor and store it in an airtight bottle. You can, however, varnish the specimen. Although this spoils the appearance it does halt the decay.

Pebble polishing is a popular hobby. Stones are placed in a revolving drum with abrasive powder, and left to churn about for several weeks. The highly-polished pebbles that result are very attractive, and may show off features of their constituents that would otherwise be obscure. The different grain sizes in a porphyry, and the contrast between the white calcite of a fossil and the black calcite of its limestone matrix can be seen after polishing. Hand specimens with faces polished on a grinding wheel, or lap, will also show these features. Do not be tempted to varnish the polished surface. You will lose all its natural beauty that way.

Some collectors embed small specimens in casting resin, producing solid transparent blocks with a mineral crystal inside. This technique is very successful with biological specimens – in fact it is often necessary to prevent the organic material from decaying – but it is less suitable for geological material. The natural beauty and impression of robustness of the specimen is lost by such treatment.

Arrangements

Geologists tend to be specialists. This is as true of the amateur as it is of the professional. You may have a particular interest in the different forms of granite, or the sequences of rocks and fossils of the Jurassic period, or the minerals of metamorphic rocks. If so, let your collection reflect this.

Remember, someday your collection may have grown so much that it may be the most comprehensive one of its kind in your particular field. It may be of great importance to the study of the subject. Make your preparations with the possibility of such immortality in mind.

and the pea-sized garnets nestling in their augens; your perfectly transparent quartz crystal, as big as a walnut, extracted carefully from the cavity in the vein on the granite tor; your beautifully preserved ammonite fossil with all the chambers in its shell visible – all these deserve special treatment. You will want to show them off.

A glass case, or one of transparent plastic, is best. The exhibit can be mounted on black velvet or colored satin or felt, or whatever is aesthetically appropriate. Carefully angled lighting can show off its best features. A mirror can be incorporated into the back of the case if there is anything interesting to see at the back of the specimen.

Although rock samples tend to be quite strong, there are certain items that need treatment. These include fossils embedded in shale which may be

PREPARING REPORTS

The climax of all geologists' work is the preparation of their reports. In these they present to the world what they have been doing, how they have done it, and what results they have obtained. The report must be detailed and comprehensive. It must also be readable by the people who will use it, be they other geologists, university professors, quarry owners, or drilling engineers.

Abstract

The abstract is the summary. It always precedes the report itself, so that anyone studying the subject needs only to look at it to see if the report is going to contain anything relevant to their particular work. It should only be a paragraph long. Do not write:

> "The geology of the Zedd Valley is studied, the outcrops measured and analyzed, the sequence of rock types described and the minerals identified. The orientations of lineations are noted."

because this is what the report is expected to do anyway. Instead, say something like:

> "The Zedd Valley is cut through slates, which consist of metamorphosed Ordovician shales. In Doemann's Quarry, relict bedding was seen with distorted graptolite fossils. The metamorphic grade was quite light over most of the area but reached garnet grade in the northwest, where well-formed garnets were seen and collected. Orientation of the cleavage varied from 35° to 50° and dipped at an angle of 70° to 80° toward the southeast . . ."

This approach deals with your results rather than your procedure, which is what the reader wants to know in the first place. A good hint is to write your abstract last with the intention being to give an overview of the paper's subject and conclusions.

Text

The main text of your report should first of all establish the location and then cover the actual work. It is not necessary to deal with the sites in the order in which they were visited. It may be more sensible to take the oldest rocks first and deal with them wherever they appear, and then the whole succession chronologically.

Samples should be described in as much detail as possible, and reference made to how they can be found in your collection.

Illustrations

The best reports tend to be well illustrated.

Start with a sketch map of your study area. If your area is likely to be remote and unfamiliar, make it a map within a small-scale map of a bigger region.

Present your photographs with annotation. Either write on the photographs themselves with a suitable pen, or tape a tracing paper or plastic overlay over them and indicate anything that needs to be highlighted.

Sedimentary sequences are usually shown as

Right Your geological report begins life as a notebook full of field notes and drawings such as these. Any laboratory work is then based on the specimens brought home with you, and on the information in the notes. The final report will bring together the field observations, the laboratory studies, and the conclusions drawn from the two.

columns, using the conventional symbols to identify the rock types (see p. 65) and with each bed drawn to scale. It is quite satisfying when, having found the same sequence of beds at different outcrops, you can draw several columns side by side and correlate particular beds with dotted lines drawn between them.

The map with its cross section is usually the illustrative highlight of the report, often presented as a separate, large foldout map at the back of the volume. The beds that outcrop can be colored with felt-tip pen, colored pencil, or watercolor, as can the inferred geology beneath the drift between. However, it must be made clear what represents outcrop and what is inferred.

Conclusion

The conclusion may read something like the abstract. If you had set out to prove something, say if you proved it or not. If the results are inconclusive, present them as fully as possible so that your report can be of use to anyone who will be following in your footsteps to carry out similar studies in the same area.

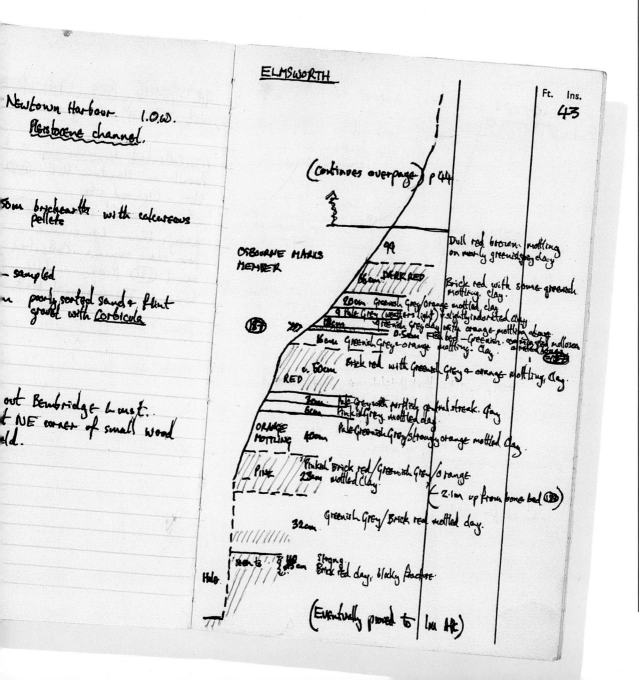

GEOLOGICAL SITES

It is difficult to give a comprehensive geological guide to a particular area – impossible with an area the size of a continent. However, a brief account of the geological history of each continent will give the practical geologist some idea as to where the particular rocks occur and why they are there. The following is as close as we can get to a geological guide to the world – in a dozen pages!

AUSTRALIA

Australia's oldest rocks lie in the Yilgarn Shield in the extreme west of the continent. The mineral deposits of these ancient rocks give rise to the Coolgardie gold mines. Late Precambrian rocks in the Adelaide trough consist of shallow water deposits such as sandstones. They contain the Ediacara fauna – a unique suite of Precambrian fossils consisting of worm-like, seapen-like, and jellyfish-like creatures.

The eastern mountains grow

This region, near Adelaide, represented the eastern boundary of Australia, in fact the eastern boundary of Gondwana. Anything farther east from that was deep sea, and in Cambrian times a subducting plate margin developed here. The normal process of continental accretion took place here – island arcs and volcanoes, followed by continental edge mountain ranges, which then began to erode and were replaced by new ranges at the seaward side. The mountain-building activity shifted from the Adelaide region in the early Paleozoic toward the present-day coastline as the Paleozoic ended.

As the mountains were rising in the east, the majority of the continent remained above sea level. Huge areas of inland deposits in the Cambrian produced sandstones about 4 miles (6 km) thick. The only remnants of these are the spectacular outcrops of Ayers Rock and the Olgas in the Northern Territories. The northern and western edges of the continent were periodically flooded by shallow seas. In the Devonian rocks of northern Australia lies the Canning Basin. Here is a mass of limestone, 220 miles (350 km) long, that represents a Great Barrier Reef. This is similar to the modern one along the east coast, except that it was formed from algae rather than corals.

Many parts of the south have tillites – the Marinoan and Sturtian tillites – dating from a Permian ice age.

The flooding of the continent

The continent remained largely above water until the Cretaceous. The area known as the Great Artesian Basin had continental deposition in Jurassic times, when lakes and coal swamps spread across the interior. Early in Cretaceous times the sea began to flood the continent, and so almost a third of the Australian land surface consists of Cretaceous rocks. At the end of the period the sea withdrew, and there were again river deposits formed in the Great Artesian Basin.

The breakup with Antarctica

In the early Tertiary, rifts appeared between Australia and Antarctica. The whole Australian continent then drifted away from what was left of Gondwana and began moving northward. The cause of the movement was the northward movement of the Australian Plate. This plate is being destroyed along the Java Trench and the Australian continent has almost reached there. During the Tertiary, Australia was nearly all dry land. All this time the continent and its animal life have been isolated from all other landmasses, resulting in the particular fauna of marsupials that now exists there.

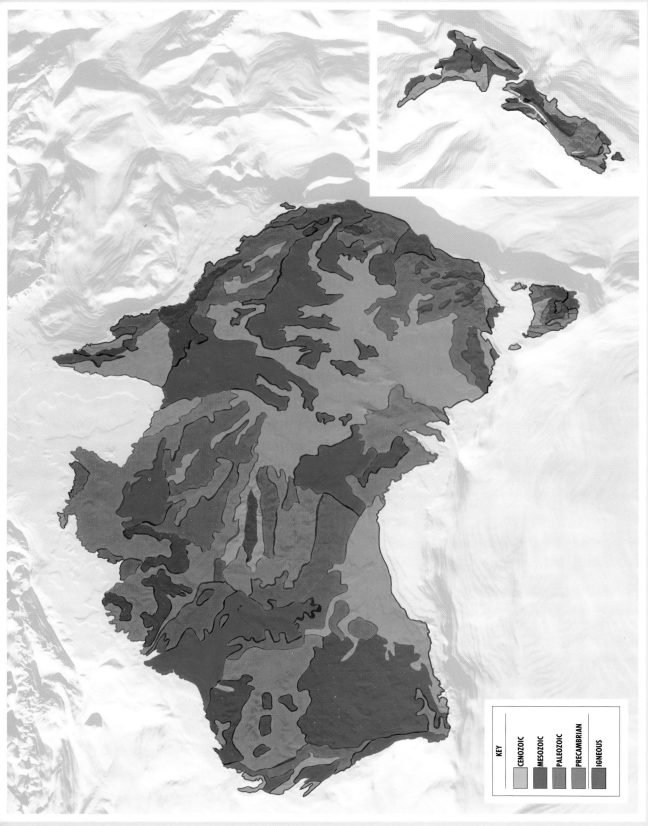

KEY

CENOZOIC

MESOZOIC

PALEOZOIC

PRECAMBRIAN

IGNEOUS

ASIA

The vast continent of modern Asia is to some degree an amalgam also. The main portion of its basement is the Siberian Shield that occupies the low-lying regions of the north. The other areas of basement lie in China and in Kazakhstan, which, at the beginning of the Cambrian, were separate landmasses, and in India which was buried deep in Gondwana. Most of these were strung along the equator, and so Cambrian deposits tend to be of tropical limestones. It is not clear just when these various pieces were combined to form the modern continent of Asia. India was part of Gondwana until very late in history, but the others may have been together since the early Paleozoic, or they may have remained separate until late Carboniferous times when Siberia fused with Europe and they all merged together as part of Pangaea.

Northeastern Pangaea

The whole landmass of Europe and Asia constituted the northeastern part of Pangaea, separated from the southern part, Gondwana, by the vast embayment of the Tethys Sea. Fold mountains lay along the Tethys shore of southern Asia and the remains of these are now found in the central Asian mountains, to the north of the Gobi Desert. The plate margin of Southeast Asia has a complex history, with islands appearing and disappearing in the region of Japan and Korea from the Triassic onward.

In the Jurassic the sea levels rose, and covered the area of northern Siberia. Huge tracts of China, Malaysia, and Indonesia were left upstanding as a dry continent. As the Jurassic developed, the shallow northern sea spread southward in the region of the Ob river. At one time the sea became so extensive that the Tethys was continuous with the contemporary Arctic Ocean. Uplift in the Cretaceous meant that most of Asia was dry land from that time onward.

During the Cretaceous the ancient Pangaean landmass fragmented, and today's continents drifted apart. This led to an increase in the activity of the destructive plate margins (subduction zones) along the east coast of the continent – an activity that is continuing until the present day in the volcano and earthquake activity of Japan, the Philippines, and New Guinea.

The impact of India

As Gondwana broke up, a triangular mass broke away from between Africa and Antarctica. This became the subcontinent of India. In one of the most spectacular cases of continental drift, this subcontinent departed from its parent landmass, swept across the Tethys Sea and collided with the southern edge of Asia. A destructive plate margin along this southern edge drew the continent northward, producing the usual pattern of ocean troughs and volcanic island arcs as it did so. When the collision occurred in the early Tertiary, these deep-sea sediments and volcanics were uplifted as the Himalayan mountain range. The eastern portion of this destructive plate margin is still active, producing the ocean troughs and volcanic island arcs of the East Indies. During its movement, India passed over a spot of intense magmatic convection deep in the mantle, like that which causes the eruptions in Hawaii today. The result was a vast sequence of basaltic lavas that now form the Deccan Plateau in the west of the subcontinent.

The loss of the Tethys

All these factors led to the closure of the Tethys Sea. Africa also drifted northward, colliding with Europe, and the Saudi Arabian plate collided with Iran producing its northwest-southeast mountain ranges. The Tethys was squeezed out of existence, leaving only a remnant sea, the Paratethys and the paleo-Mediterranean. The Paratethys itself eventually became what are now the Black Sea, Caspian Sea, and Aral Sea.

Meanwhile, during the Tertiary the sea came and went over the low-lying lands of the far north. The arm of the sea that reached along the basin of the Ob, joining the northern and the southern oceans, opened and closed several times.

Because of problems in dating the rocks of the Himalayas, however, it will be some time before we have a clear chronology of this area's history.

The start of a new rift

As a final indication that continental movement is still taking place, Lake Baikal represents an embryo ocean in a rift valley. This may mean that the far eastern portion of Asia may eventually fragment into a new continent.

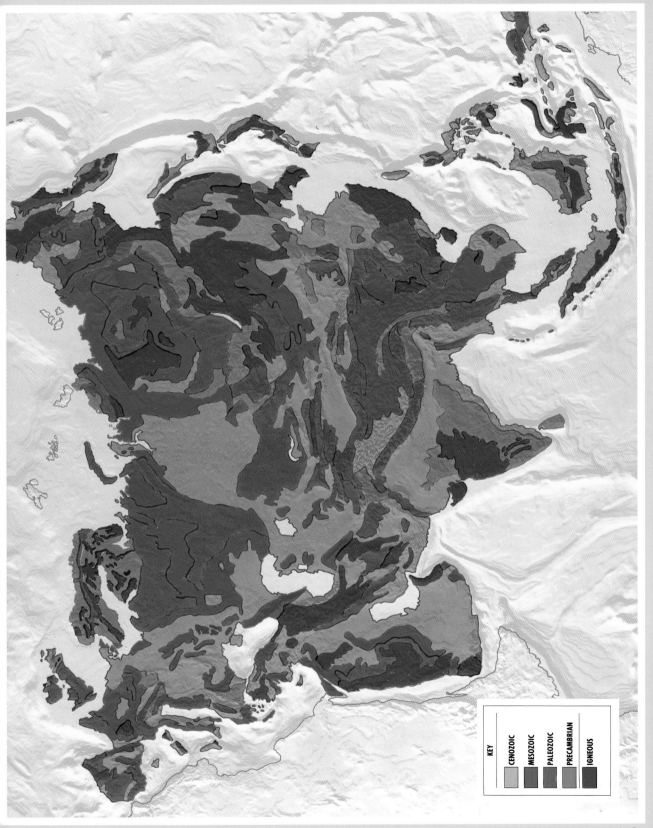

KEY

CENOZOIC MESOZOIC PALEOZOIC PRECAMBRIAN IGNEOUS

EUROPE

Europe's geology is far more complex than that of Africa and South America, because of a number of separate tectonic collisions which severely distort the geological beds. Europe consists of Precambrian basement fragments consolidated from numerous sources, surrounded by sediments from every period of geological time, which in turn have been twisted into mountain chains and cooked into metamorphic masses.

The main Precambrian block is the Baltic Shield, now exposed in the low-lying areas of Scandinavia but covered by later sedimentary rocks to the south and east. Another piece of Precambrian now lies in northwest Scotland, and this started off as part of the Canadian Shield with a thick covering of desert sandstones. The bits and pieces that were to become southern Europe were probably part of Gondwana.

The collision with America

The first post-Precambrian event that began to shape Europe as we know it was the collision of the North American continent with that based on the Baltic Shield. In Cambrian, Ordovician, and Silurian times the intervening ocean closed, and the shales and limestones formed there began to be squeezed up in between. The continents collided in the Devonian, folding these sediments into a vast mountain range that stretched from the Appalachians, through northern Scotland and across Norway. Its remains are still seen in the highlands of these areas.

Immediately these mountains began to erode, spreading river- and desert-sands southward in extensive sequences known as the Old Red Sandstone, now found in Scotland and Poland. In the south these beds were flooded by shallow seas, giving deposits of Devonian limestone in Germany.

Gondwana joins in

Then, in the Carboniferous, this "super" continent collided with Gondwana and the areas that were to become southern Europe were annexed, the joint sealed by mountains whose deep batholiths now lie in northwestern France, southwestern England, the western half of Spain and Portugal, and southern Germany. Between these mountains, and the older ones farther north, there spread shallow seas that were then encroached upon by vast sweeps of swampy forested delta. The results are early Carboniferous limestones, followed by the late Carboniferous coalfields of Britain, Germany, and Poland.

Asia collides

Then, in the Permian, yet another continent approached and collided with the eastern margin, uplifting the Ural Mountains. It was a time of deserts, and desert sandstones were deposited, although shallow seas left their mark as limestone, dolomite, and evaporite beds in Germany and northeastern England. These dry conditions continued into the Triassic, and desert sandstones – the New Red Sandstones – attest to the fact in many areas.

Although all the continents were, by now, fused into a single supercontinent called Pangaea, there was an ocean that formed a vast embayment between eastern Gondwana and the European-Asian section in the north. In the shallows of the northern edge of this ocean – the Tethys Sea – lay reefs and islands in the region of southern Europe. Muds from the Tethys lie as shales in northern Italy and the Carpathians. The Tethys also spread over the rest of Europe, giving the Jurassic limestones of England and central Europe. The shallows of Bavaria produced a limestone so fine that even the feathers of Jurassic birds are fossilized in it.

As the continent of North America tore itself away in the Cretaceous, shallow seas spread over most of Europe, depositing vast layers of chalk which are now so characteristic of the landscape of southern England and northern France. The rifting of the North Atlantic spread basalt plateaus that now cover Northern Ireland and the Scottish islands.

The Mediterranean – the buffer with Africa

The Tethys closed during the first half of the Tertiary. Africa collided with southern Europe, dragging its edge toward the east. Spain rotated counterclockwise. Corsica and Sardinia were plucked from the French landmass. The Atlas Mountains in north Africa and the Alps were thrust up, and then twisted into a great S-shape, with Sicily and Italy between them.

Finally the ice age of the last two million years gouged out mountain valleys and spread glacial deposits over much of the northern lowlands.

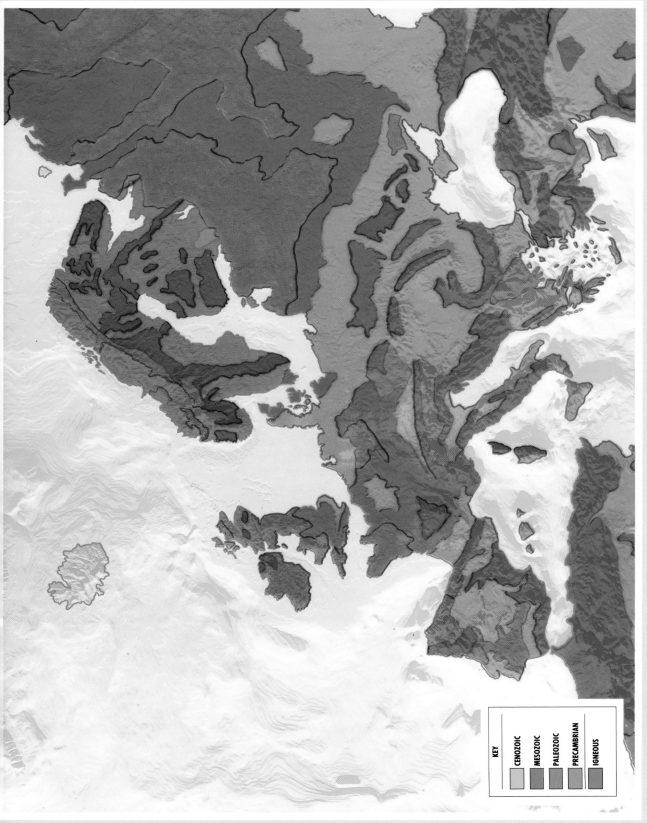

KEY

CENOZOIC

MESOZOIC

PALEOZOIC

PRECAMBRIAN

IGNEOUS

AFRICA

Africa is what remains of central Gondwana. It is built around three Precambrian shields. These are the Mauretanian Shield in the northwestern lobe (geologically continuous with the Guyana Shield of South America), the Congo Shield of Central Africa, and the Kalahari Shield of the south. Apart from some fold mountains in the extreme northwest and the very south, and the brand-new rift system of the east, the continent has been relatively stable for its whole history.

While it was part of Gondwana the only part of the continent open to the sea was the extreme north. This was the edge of a broad ocean that geologists call the Tethys Sea which separated Gondwana from the continents of the north. The southern tip was also exposed and from time to time became part of the Samfrau mobile belt.

Early ice ages
The glaciation that occurred in South America in Ordovician times also affected the northern parts of Africa. Tillites can be found in the northwest of the Sahara Desert and in the southwestern corner of the Arabian Peninsula. Through most of Gondwana's history the African region was above water, and so there is little in the way of post-Precambrian sedimentary rocks on the continent.

North America collided and fused with the northwestern corner of Africa in Carboniferous times, uplifting the mountain range that we now see as the western part of the Atlas Mountains. They were then continuous with the Appalachians. Sediment that originated from these mountains produced fertile soil and coal swamps in Algeria. Further east ocean water flooded the northern edge of the continent, and dried out to give vast salt deposits (gypsum) that now lie in the middle of the Sahara Desert.

More tillites, from the Permo-Carboniferous glaciation, lie in the southern half of the continent. Termed the Dwyka tillite, it outcrops extensively in South Africa, where most of it is overlain by later rocks, and in Uganda.

In the southern part of the continent these tillites are followed by the Karroo System, an extensive series of Permian, Triassic, and Jurassic lake deposits, delta sandstones, and desert sediments. The lake deposits contain fossils of the swimming reptile *Mesosaurus*, found also in South America.

There are also fossils of the mammal-like reptile *Lystrosaurus*, also found in India and Antarctica – further proof of drifting continents.

A time of stress
At the top of the Karroo lies a sequence of basalts, sometimes almost a mile (over a kilometer) thick. These now form the Drakensberg Mountains and other prominent highlands, and erupted from volcanoes that accompanied the formation of a rift that would eventually separate Africa from Antarctica as Gondwana broke up.

This fragmentation did not take place until Cretaceous times, when South America drifted toward the west as well. If you take a thin slab of clay and push it up from beneath it usually splits into three radiating cracks. This tends to be the pattern of rift valleys formed when continents break up. In the case of South America and Africa, one rift reached southward from the area of modern Nigeria, one reached westward, and together these two formed the modern continental margins. The third, a failed rift, reached northeastward and is still present as a line of volcanoes through Cameroon.

During the late Cretaceous there was a rise in global sea levels. Most of the north was flooded and an arm of the sea – the Trans-Saharan Seaway – reached across the continent, cutting off the western lobe. This survived until Tertiary times, by which time the Tethys Ocean had closed and Europe was grinding against the northern coastline. The remainder of the Atlas Mountains were uplifted, and this activity is still underway.

In the early Tertiary, new rifting commenced along the eastern margin of the continent. Another triple rift appeared in the region of Ethiopia. One arm formed the Red Sea, the second extended into the Indian Ocean, and the third stretched southward forming the Great Rift Valley systems of East Africa. Time will tell whether or not this is yet another failed rift, or if a new continent will eventually shear off into the Indian Ocean.

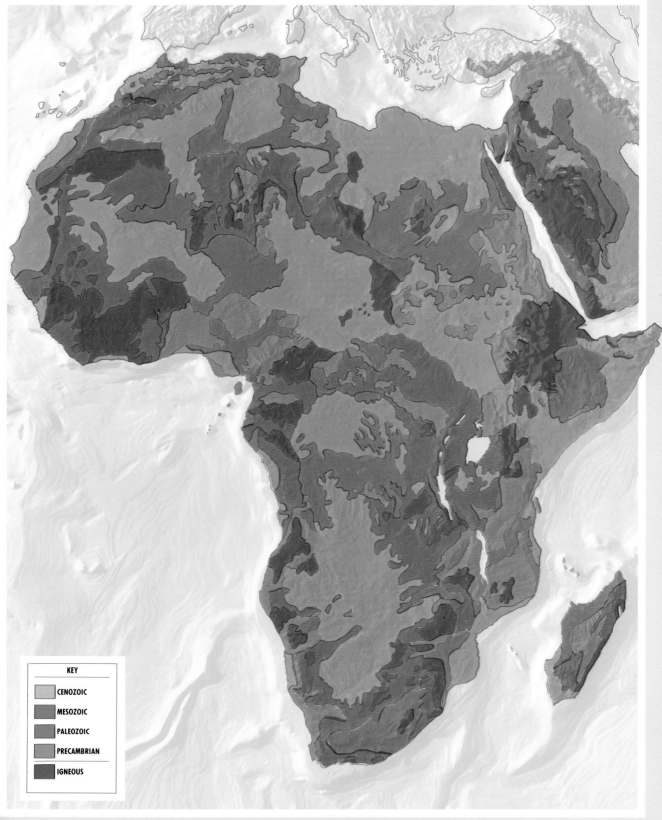

KEY

CENOZOIC

MESOZOIC

PALEOZOIC

PRECAMBRIAN

IGNEOUS

SOUTH AMERICA

Until the late Cretaceous, South America was merely the westernmost portion of the great southern supercontinent of Gondwana. We can tell this by the most obvious fact that the eastern coastline seems to fit closely with the western coastline of Africa, and that many of the pre-Cretaceous rock formations and mountain chains are continuous across the intervening South Atlantic ocean. As with other continents, the Precambrian shields form the core of South America, although they are detached along the rifts of the Atlantic coast. Two main shields are present – the Guyana Shield in the north, and the Brazilian Shield that occupies most of the broadest part of the continent. Three smaller shields are scattered down the tail.

The oldest active mountain chain

The South American west coast is dominated by the Andes, the longest uninterrupted mountain chain in the world. This has its origins as the western coastline of Gondwana throughout most of its history. A destructive plate margin (subduction zone) was producing ocean troughs and volcanic island arcs here in early Paleozoic times. The destructive margin was actually continuous along the entire coast of Gondwana, from South America, through Africa and Antarctica to Australia and New Guinea. It is often referred to as the Samfrau mobile belt, a name derived from South AMerica aFRica AUstralia.

In early Paleozoic times this corner of Gondwana drifted from equatorial latitudes down toward the South Pole. By the late Ordovician and early Silurian there was an ice age in the southern continents. The boulder clay – that mixture of clay and rubble left behind by the glaciers (pp. 80–81) – solidified to form the rock called tillite, and tillites dating from Silurian times outcrop in Argentina and Bolivia.

Throughout most of its history the shield areas were above the water, and so there were few sedimentary rocks deposited upon them. Between the Guyana and Brazilian Shields there were periodic transgressions of the sea during the Paleozoic. The result is a sequence of sedimentary rocks less robust than the flanking shields, and so the Amazon River eroded its course through these.

Proof of Gondwana

Toward the end of the Paleozoic – in Carboniferous and Permian times – the continent was close to the South Pole. Another ice age ensued and the tillites from this are found on the shield areas in the south of the continent. Fossil striations and other ice-formed features show that the ice came over from the direction of Africa, and match up with the glaciated features on that continent. This was one of the first indications that continental drift had occurred.

Another such indication appeared in the Permian. After the ice age came a more temperate period producing coal forests in Brazil. Freshwater shales deposited at this time have the fossils of a freshwater swimming reptile called *Mesosaurus*. The exact same animal is found in similar shales in South Africa, now at the other side of the Atlantic. By this time Gondwana had merged with the northern continents to become Pangaea, and North America lay close by to the north.

The Andes were established by this time; eruptions of typical andesitic volcanoes were frequent and sediments from the mountains were being deposited beyond their foothills in the continent's interior. The deep magma chambers that supplied the molten material for the Mesozoic volcanoes are now exposed as granite batholiths along the length of the range. The "tail" of South America began to develop. A broad swathe of Jurassic rock lies to the east of the Andes from Bolivia southward.

The supercontinents break up

In the late Jurassic, South America rafted away from North America, shearing off pieces of continental crust that eventually became Central America and the West Indies. It did not separate from Africa until the late Cretaceous. We can tell this because the dinosaur fauna suddenly became very different from that of elsewhere, just as modern Australia's isolated fauna is different from that of the rest of the world.

During Tertiary times South America was still an island, and the land bridge to North America was not completed until about 3 million years ago. During the last ice age South America suffered more than the other southern continents. Ice caps and glaciers spread all over the southern Andes, leaving their debris on the surrounding lowlands.

KEY

CENOZOIC

MESOZOIC

PALEOZOIC

PRECAMBRIAN

IGNEOUS

NORTH AMERICA

The North American continent conforms quite nicely to the idealized structure of a continent. The central portion is a shield – a flat area of metamorphosed basement rock dating from Precambrian times. Some of it is exposed, but some of it is covered with later sedimentary rocks. Around it are mountain chains that become younger toward the coast. How did it get to be this way?

The ancient heart

At the end of the Precambrian, the continent consisted of little more than the Canadian Shield that now lies exposed over much of Canada and underlies the sediments of the midwestern United States. To the northeast were broad continental shelves, part of which, beyond Greenland, now lies in northern Europe, but we shall come to that later.

As the Cambrian began, shallow seas spread across the edge of the shield, initiating coastal sand deposition, mud deposition closely offshore, and in deeper waters, carbonate (limestone) deposition. Where the shelf fell away into the deep ocean, shales were formed. These include the famous Burgess Shale of British Columbia that contains a clear suite of Cambrian fossils.

Destructive plate margins were established along the west and southeast coasts of the continent as the Ordovician and Silurian progressed. Mountains and island arcs developed along these coasts while nearly the whole of the interior of the continent was covered with shallow water. Rocks of these times include sandstones in the Appalachian mountains that began to rise then, and limestones in the Great Lakes area.

The continents collide

The significant event of the Devonian was the collision of North America with northern Europe (the Scandinavian Shield), uplifting a Himalayan-type mountain chain along the collision site. The remains of this mountain chain lie in the northern Appalachians, and across the Atlantic in Scotland and Norway. Sands eroded from this mountain chain form the Old Red Sandstone of New York, Pennsylvania, and West Virginia. Farther west the shallow seas continued over the Canadian Shield, producing limestones in many areas, such as the Alexandra reef in Alberta, and the Chattanooga Shale in Tennessee.

Come the Carboniferous, the combined North American and northern European landmass collided with that of the combined southern continents – called Gondwana – to form the supercontinent of Pangaea, and raised the rest of the Appalachians. Swamps along the foothills produced the Pennsylvania coal deposits. Limestones that had been forming in the shallow waters over much of the remainder of the continent include the Redwall limestone in the Grand Canyon and the Madison farther north. In the far west a mountain range was rising along the edge of the continental mass – the ancestral Rockies.

In the succeeding Permian, the sea retreated from much of the continent and deserts prevailed. The sea became restricted to the far south, and the reef deposits formed then are known from the Guadalupe Mountains in Texas and New Mexico.

The mountains between the North American and the European continents continued to erode in Triassic times, producing the brownstone sandstones of Connecticut. The rest of the continent was dry land, and the flora is preserved as the Petrified Forest of Arizona. Slightly later is the Navajo desert sandstone of Zion National Park. Along the west coast the development of the mountain range was similar to that of the modern Andes.

In the Jurassic, the sea transgressed a depression in the western mountains and flooded much of the Midwest. Riverine deposits between this and the western mountains produced the Morrison Formation that stretches from New Mexico to Montana along the foothills of the modern Rockies, famous for its wealth of dinosaur fossils.

The rift with Europe

The dramatic geographic event of the Cretaceous was the fragmentation of the supercontinent, when rifts appeared between North America and the landmass of Africa and Europe, and North America rafted away westward leaving the Atlantic Ocean between. The Rockies were by now quite well established, and new mountains that were to become the Coast Ranges began to rise farther west. A vast area of shallow sea spread southward from the Arctic and formed the deposits of chalk, particularly in Kansas. Limestones were deposited in the south and along the new eastern coastline.

Tertiary and Quaternary times bring us up to the present day, with continuing uplift in the west.

KEY

CENOZOIC

MESOZOIC

PALEOZOIC

PRECAMBRIAN

IGNEOUS

THE GEOLOGICAL TIME SCALE

The 19th-century and early 20th-century estimates of the age of the Earth were based on such things as the length of time it would take the mass of the Earth to cool to its present state. More modern estimates of time

PRECAMBRIAN
Up to 570 million years ago. The first 85 percent of the Earth's history. Divided into the Archean, in which there is no sign of life, and the Proterozoic, the time of the most

primitive life. Archean rocks are mostly metamorphic. Proterozoic rocks are sedimentary, igneous, and metamorphic. "Banded iron formations" are important – showing sedimentary deposition under conditions of low atmospheric oxygen levels. Remains of two major ice ages found.

Precambrian
More than 570 million years ago

Cambrian
570–510 million years ago

Ordovician
510–439 million years ago

Silurian
439–409 million years ago

Devonian
409–363 million years ago

Carboniferous
363–290 million years ago

PALEOZOIC
Cambrian 570 to 510 million years ago. The time of the first good fossils. North America separate from Europe. Gondwana lying in the southern hemisphere. Widespread flooding of continents, producing large areas of limestones. Shales and sandstones also common. Atmospheric oxygen levels rising.

Ordovician 510 to 439 million years ago. Seas retreated from the continents and then transgressed them again. More shallow marine sediments – undisturbed over shield areas or caught up in mountain-building. Deep-sea sediments in

western North America and Europe where the continents were moving together. Continents generally moving southward.

Silurian 439 to 409 million years ago. Another ice age at the beginning of this period. More flooding and drying out of the continents. Many limestones and mudstones. Fossils of the earliest land plants found. North America closing with Europe, squeezing up mountains (the Caledonian Mountains) in between. Gondwana covers the south polar region.

Devonian 409 to 363 million years ago. Continents moving together, producing large continents with desert sandstones and river deposits in many areas – the Old Red Sandstone.

are based on radioactivity. Radioactive elements decay at known rates from one form, or isotope, to another. By measuring the relative abundances of the radioactive isotopes and their "daughter isotopes" in a particular rock, geologists can tell how long ago the rock formed.

For example, half of any mass of uranium-238 (the number refers to the number of particles in the nucleus of the atom) decays to the daughter, lead-206, in 4510 million years. This is known as the isotope's "half-life." The half-life of thorium-232 as it decays to lead-208 is 13,900 million years. These can both be used to date rocks several tens of millions of years old. Carbon-14, however, as it decays to nitrogen-14, has a half-life of only 5570 years, and so it is only of use in dating recent material such as human artifacts.

MESOZOIC

Triassic 245 to 208 million years ago. Still desert sandstones. Complete change in animal fossils from the end of the Permian. Final few fragments of continent united with main supercontinent to complete Pangaea. Volcanic activity produced vast quantities of plateau basalts in Siberia and South Africa.

Jurassic 208 to 146 million years ago. Single supercontinent of Pangaea began to fragment into separate continents, starting with the split of North America from Africa and Europe. Shallow seas transgressed the broken edges. Shales, clays, limestones, and sandstones formed. Continental deposits show moist swampy conditions, where dinosaurs flourished.

Cretaceous 146 to 65 million years ago. Supercontinent almost totally broken up. Few marine sediments from early part of the period, but extensive areas, particularly of chalk, toward the end, as the seas flooded the continental edges. Climates equable and mild.

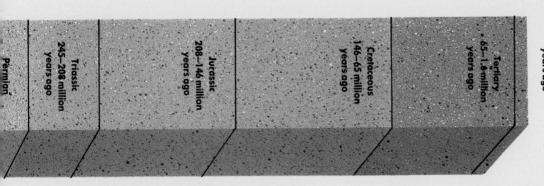

Permian | Triassic 245–208 million years ago | Jurassic 208–146 million years ago | Cretaceous 146–65 million years ago | Tertiary 65–1.6 million years ago | Quaternary 1.6 million years ago

Europe and North America in southern tropical belt. The first good evidence of land animals in continental rocks. Variety of marine sediments around the edges of the continents.

Carboniferous 363 to 290 million years ago. Regarded as two intervals in the U.S. – the Mississippian – 363 to 323 million years ago – and the Pennsylvanian. Continents still moving together. Gondwana moves toward Europe forming mountains (the Hercynian Mountains) in between. Early part characterized by shallow-water limestones; latter part characterized by delta sediments and coal measures. North America and Europe on the equator giving tropical conditions. Ice age in the southern hemisphere at the end of this period.

Permian 290 to 245 million years ago. Asia collided with Europe, producing Ural Mountains. Seas withdrew from the supercontinent again, giving a time of deposition of desert sandstones – the New Red Sandstone.

CENOZOIC

Tertiary 65 to 1.6 million years ago. Modern Earth history. Most sediments are unconsolidated, consisting of sands and muds that have not been turned to solid rock. Complete change in fossil record once more, with animal and plant life becoming similar to that of the present day. Vast deposits of basalt associated with opening of the Atlantic Ocean. South America isolated from North – Andes forming along the spine. India broke away from Gondwana and collided with Asia, producing Himalayas. Mediterranean and Alps formed as Africa collided with Europe.

Quaternary 1.6 million years ago to the present day. This is the time of the great ice ages and the development of mild interglacial ages. Most significant rock-types deposited are the tills and boulder clays associated with the movement of glaciers.

GLOSSARY

acidic: of an igneous rock, one that contains more than 66 percent silica.

anomaly: any reading that is different from what is expected. For example, a gravity anomaly at a particular place will indicate that there is a rock of a different density underground.

anticline: a fold that arches upward.

attrition: the process by which rocks and sand grains carried along by a current are worn away.

auriole: the region surrounding an igneous rock in which the native rocks have been affected by thermal metamorphism.

axial plane: the plane that joins the crests of successive beds in a fold, so that the beds on each side are more or less symmetrical.

basement: a mass of metamorphic rock that has been exposed once the overlying mountains have been worn away.

basic: of an igneous rock, one that contains between 45 and 52 percent silica.

basin: a geological structure in which beds dip inward from all directions. The opposite of a dome.

Becke line: in microscopic mineralogy, a line of light around the margin of a mineral concentrated by the refraction of light passing through the mineral.

biogenic: of a sedimentary rock, one formed from particles built up by living matter – e.g. shelly limestone or coal.

blastoporphyritic: term applied when a porphyritic igneous rock is metamorphosed, but the original porphyritic texture is still visible.

chemical: of a sedimentary rock, one built up from material deposited from solution, such as rock salt.

clastic: of a sedimentary rock, one built up from the broken fragments of pre-existing rocks, such as sandstone.

cleavage: the tendency for a mineral or a rock to split in a certain direction.

clint: in the chemical weathering of limestone, an upstanding mass of stone surrounded by dissolved hollows, or grikes.

competent bed: a bed that breaks up in a brittle manner, rather than bending, when deformed by folding.

contact: of a metamorphic rock, one that is altered by heat – through contact with an igneous mass.

continental drift: the theory that the continents have been moving throughout their history and have not always occupied the places they occupy today. The concept is now explained by plate tectonics.

corrasion: the erosive force exerted by stones and rubble dragged along a river or sea bed.

cyclical sequence: a series of sedimentary beds that repeat themselves, showing the recurring nature of the events that formed them.

delta: an area at the mouth of a river consisting of channels and islands built up from the deposition of material carried along and deposited by the river.

dendritic: anything that has a branching shape – for example, a river pattern or a crystal growth.

dike: a structure of igneous rock formed as molten material squeezes through a crack and solidifies.

dip: the slope of a bed of rock. It is the direction in which water will run when poured on the bed. The angle of dip is the angle that this makes with the horizontal.

dome: a geological structure in which the beds all dip away from a central point. The opposite of a basin.

drift: of a map, one that shows the nature of the soil and loose material lying upon the solid rocks.

earthquake: the sudden, sometimes destructive, agitation of the Earth's surface associated with the movement of rocks along a fault.

facies: the total character of a rock or a series of rocks, including the mineral composition, the structures, the fossils, and any other relevant feature.

fault: a crack in the rocks along which the rocks have moved in relation to one another.

flute casts: raised structures on the underside of a sedimentary bed formed by the infilling of hollows scooped out of the underlying sediment by seafloor currents.

foliation: the arrangement of minerals, particularly in a metamorphic rock, in distinct bands or layers. In the U.S. the term is used for the splitting of rock along the cleavage plane that is typical of schist (elsewhere termed schistosity).

fracture: in mineralogy, the nature of the surface left behind when a piece of a mineral is broken.

fractionation: the separating of different minerals at different stages of the cooling and solidifying of an igneous mass.

geomorphology: the study of landforms and landscapes.

geophone: a device for detecting vibrations transmitted through the Earth's crust.

Gondwana: the name given to the supercontinent that once included all of the continents of the modern southern hemisphere. Often called Gondwanaland.

graben: a structure consisting of a block subsided between two faults.

gravimeter: a device for measuring gravity anomalies.

greenhouse effect: the effect of the buildup of such gases as carbon dioxide, methane, and water vapor in the atmosphere, giving rise to increased ground temperatures and general climatic changes.

groundmass: in mineralogy, the substance of the rock in which a particular mineral is embedded.

groundwater: water seeping through the ground.

grike: a crack in exposed limestone that has been widened by solution.

hade: the angle that a fault makes measured against the vertical.

horst: the opposite of a graben, a block left upstanding while blocks at each side have been downfaulted.

imbricate: a structure in which flattened stones are deposited leaning on one another, like books falling over on a bookshelf.

incompetent bed: one that deforms and bends easily when acted upon by pressure.

inselberg: a rounded mass of rock, shaped by exfoliation, often forming a prominent landscape feature in arid areas; also called a kopje.

joint: a crack in the rock which, unlike a fault, shows no movement of the rocks themselves.

lagoon: an area of shallow water cut off from the sea by a sand bar or by a reef.

lateral fault: a fault in which the movement of the rocks has been horizontal.

Laurasia: the supercontinent that comprised what are now North America, Europe, and Asia.

levée: the bank along the

side of a river, formed by the deposition of river material at times of flood.

lithification: the process by which a loose sediment becomes a sedimentary rock.

luster: in mineralogy, the way a mineral catches the light and reflects it, used particularly as an aid to identification.

massive: applied to a sedimentary rock, one that shows no bedding.

meteorite: a mass of interplanetary matter that has fallen to the surface of the Earth.

mullions: cylindrical structures formed in metamorphic rock as competent beds are broken up and ground against one another.

neck: a cylindrical igneous structure formed as molten material solidifies in the vent of a volcano.

nodule: a chunk of mineral, usually of irregular shape.

ocean ridge: a ridge along the ocean floor formed as molten material erupts along the edge of a tectonic plate. The plates grow from the ocean ridges.

ocean trench: an elongated trench along the edge of an ocean where one tectonic plate is being pulled down beneath another.

ore: a mineral from which a useful metal can be obtained.

paleogeography: the study of ancient landscapes, land and sea distribution, and sedimentary conditions.

paleontology: the study of fossils and the life of the past.

Pangaea: the supercontinent that once included all of the continents in a single landmass.

pegmatite: a rock found in a vein and consisting of very coarse crystals.

permafrost: the condition found in tundra areas in which the soil is permanently frozen at depth even though the surface thaws seasonally.

petrology: the study of rocks.

photomicrograph: a photograph taken through a microscope.

plumb line: a surveying device consisting of a weighted string, used to determine the vertical.

plunge: the direction and angle of the dip of the axis of a fold.

polarizing filter: an optical filter that allows only light polarized in one plane to pass through it.

porphyritic: a texture of igneous rocks in which large crystals are embedded in a finer groundmass.

puckers: small-scale folds on the limb of a larger fold.

pumice: solidified lava that is so full of gas bubbles that it floats on water.

reflection: the action of waves of light, sound, or other vibrations bouncing off a surface.

refraction: the action of waves of light, sound, or other vibrations altering direction when passing from one medium into another.

refractive index: a figure applied to a substance that shows the amount by which light will be refracted while passing through the substance. A mineral with a high refractive index will refract light more than one with a low refractive index.

regional: of metamorphic rocks, one that is altered by pressure.

regression: the withdrawal of the sea from an area, by the lowering of the sea level or raising of the land.

rift valley: a valley formed by the subsidence of a tract of land between faults.

rock: a general term for any substance that makes up the material of the Earth.

rock cycle: the process of the formation of a rock, its exposure to the forces of erosion, its breakdown, and the accumulation of its material to form new rock.

scarp: the steep slope formed as the edge of a bed outcrops – as opposed to the dip which is the slope formed by the surface of the bed.

scree: a slope of broken rocks that have been weathered off an exposed outcrop by the action of frost.

seafloor spreading: in plate tectonics, the growth of new plate material by its accretion at oceanic ridges, causing the plates at each side to move apart.

sediment: any loose material deposited by natural means.

seismic: pertaining to earthquakes.

shield: a large area of ancient metamorphic rock that forms the heart of a continent.

slickenside: polish marks along a fault plane caused by the movement of the blocks.

soil: the loose covering of the Earth's surface consisting of broken rock fragments and decaying vegetable matter.

solar wind: the constant radiation of particles from the sun.

solid: of a geological map, one that shows the rocks rather than the soil cover.

strata: beds of sedimentary rock.

stratigraphy: the study of the geological history of an area by the kinds of sedimentary rocks occurring, and the structures and fossils within them. Sometimes called historical geology.

streak: the diagnostic mark left by a mineral as it is scratched across a hard rough surface.

strike: the horizontal line made by a dipping bed.

strike line: in geological mapping, a line on a map that indicates where a particular bed would reach a particular elevation.

strike-slip: of a fault, one in which the movement has been down the slope of the fault.

subduction zone: a region at the edge of a tectonic plate where the plate is being destroyed beneath the next. This is usually marked by an oceanic trench and an island arc.

superposition: the principle that states that in any sequence of undisturbed sedimentary rock, the oldest lies at the bottom.

syncline: a fold in which the beds sag downward.

tectonic map: one that shows the geological structure of an area, rather than the rocks themselves.

Tethys: a name given to the sea that once separated the supercontinents of Gondwana and Laurasia.

thermal: another term for contact metamorphic rock.

throw: the distance that a fault has moved.

thrust: a fault caused by compression.

tor: an upstanding mass of granite, weathered into rectangular blocks by solution along joints.

trangression: the opposite of a regression – the advance of the sea over the land.

turbidity current: a cloudy mass of rock fragments and sand swept along by a water current.

ultrabasic: an igneous rock containing less that 45 per-cent silica.

vein: a crack in the rock that has filled with mineral material.

INDEX

Note: Page numbers in *italic* refer to the illustrations and captions; page numbers in **bold face** refer to main entries.

USEFUL CONTACTS

To find out more about the geology of your area or an area that you plan to visit, try the following:

* Look in the phone book for geological museums, societies, surveys, and under prospecting and mining companies.

* Visit the public library nearest to your chosen area and scan the bulletin board for events or ask for a list of local organizations. A local bookstore may also have a bulletin board, and it may stock relevant guides to the area.

* Contact the nearest university's geology department.

* Contact the local tourist board.